Reed New Holland

Text and Photographs by
Densey Clyne
Additional photography by Jim Frazier
and Glen Carruthers.

Densey Clyne's Wildlife OF AUSTRALIA

Contents

Published in Australia in 1999 by
Reed New Holland
an imprint of
New Holland Publishers (Australia) Pty Ltd
Sydney • Auckland • London • Cape Town
14 Aquatic Drive Frenchs Forest NSW 2086 Australia
218 Lake Road Northcote Auckland New Zealand
24 Nutford Place London W1H 6DQ United Kingdom
80 McKenzie Street Cape Town 8001 South Africa

First published in 1988 by Reed Books Pty Ltd
Reprinted in 1999 by New Holland Publishers (Australia) Pty Ltd

Reprinted in 1999 by New Holland Publishers (Australia) Pty Ltd

National Library of Australia Cataloguing-in-Publication data:

Clyne, Densey, 1926- .
Densey Clyne's wildlife of Australia

New ed.
Includes index.
ISBN 1 876334 31 2.

1. Zoology - Australia. 2. Zoology -Australia - Pictorial works.I. Title. II. Title: Wildlife of Australia.

590.994

Editor: James Young
Designer: Bruno Grasswill and Tricia McCallum
Cover design: Roula Doulas
Printer: Imago Productions (F.E.) Pte. Ltd.

Oecophylla smaragdina

Introduction

When I walk into a bookshop, for whatever purpose, it's natural for me to make a bee-line for the Australian wildlife section. I do this knowing that my appetite is going to be whetted far beyond the capacity of my purse or credit card.

I am dazzled afresh every time by the mouthwatering collection of new books and new editions of old ones, their dust jackets shining in rainbow display. The shelves seem to spill over with books about our wild plants and animals, begging to be bought or at least browsed through.

This was not always so. I remember a time when the average bookshop gave little shelf space to Australian natural history. In any case, new books on the subject were rare and copies of 'classics' by the early naturalists hard to find—though it was always fun searching for the occasional scoop among dusty piles of secondhand books.

Today the output of books on natural history is astonishing. They range from textbook to field guide, from diary to photo essay. The writing styles go all the way from poetically subjective to prosaically factual. There are books that delight and books that inform and some—the best—manage to do both. The fact that they sell is a sign of an upsurge of interest, not only in nature but also in reading.

It is my happy lot to live in two worlds, the world of nature and the world of words. As a writer without skill in the art of fiction, I have no worries at all about the making of plots and characters. All around me nature provides them in abundance—plots more intriguing than science fiction, characters beautiful and bizarre. And as a naturalist my pleasure lies not only in finding out how the wild world works, but also in expressing a little of my own aesthetic and emotional response to that world.

So it follows that as an avid reader of natural history books I seek in them not just the information but something of the informer. Each of us sees different things, and each of us sees the same things differently. Confined as individuals by time, space and character, we can get from books an infinitely wider range of experiences than we can ever have on our own.

For all that, there's nothing like seeing with your own eyes, and as a wildlife photographer and film maker it has been my good luck to travel much and set up camp in all sorts of wild and fascinating places.

However, lying eyeball to eyeball with a mudskipper on the tidal flats of Cape York peninsula, or having a midstream encounter with crocodiles in Arnhem Land, is not to everyone's taste. And people don't envy me those long hours of single-minded concentration, discomfort and often, let's admit it, boredom, waiting for nature's uncueable actors to do their thing in front of the camera. And hoping they won't fluff their lines . . . In any case there's no need to go to out-of-the-way places or live in discomfort to get close to nature. Everyone who's travelled along country and outback roads, even busy coastal highways, can recall with delight some brief encounter with wildlife. Camping holidays bring the exquisite

Green tree frog (Litoria caerulea).

pleasures of unfamiliar morning birdsong, evening encounters with waking nocturnal animals, and time to look more deeply into things. And for me there's as much pleasure—if not more—in following the seasonal activities of the wildlife in my suburban garden.

I hope this collection of essays about creatures large and small and mostly familiar will both inform and entertain you. But as I said before, the stories are not really mine; I have merely translated into words the plots and characters provided by nature, and for better or worse added a bit of myself.

Densey Clyne

1 Bounders of the bluebush

It was only a few years ago on a heat-hazy day in the bluebush country of outback New South Wales that I fell head over heels in love with kangaroos. Until that time I'd taken our leaping marsupials, our largest wild animals, entirely for granted. My interest lay with the insects.

It was just after sunset, clear and cloudless. I had left my van on the bumpy red-dirt road and gone on foot to track down the first cicada calling from a distant tree. And there she suddenly was, a great Eastern Grey female standing motionless a little way ahead. Below, her heavy body blended with the bluebush and porcupine grass. Above, her slender shoulders and head and high pointed ears rose in silhouette.

She could have been a pattern cut out of the pink sky. She could have been a carving in stone but her shape was more fluid, like a dark teardrop. I was struck for the first time by the simple rightness of kangaroo design.

She took off suddenly, clearing the bushes with every bound, lifted in effortless near-flight by massive thighs, tail rising and falling but never touching the ground. Then stopping, base-heavy again, part of the earth again, turning her head quite slowly, scanning for pursuit with her ears.

Now all of a sudden I saw the body proportions of those other, imported grazers and browsers—sheep, cattle and horses—as all wrong, their gait clumsy, their tails inadequate, their legs spindly. All top-weighted and unstable. I wonder how Aboriginal people felt when they saw these alien animals for the first time. Perhaps much the same as white men seeing their first macropods.

Macropod, that useful group name for all the kangaroos and wallabies, is simply Latin for 'big

JIM FRAZIER

An Eastern Grey Kangaroo (Macropus giganteus) *stands watchful against the sky.*

JIM FRAZIER

ABOVE: *A female Prettyface Wallaby* (Macropus parryi) *weighted down by a pouchful of joey.*

foot'—not to be confused with yeti, sasquatch or any other mythical monster. The name refers to the long, strong hind feet that deliver the thrust for take-off and, together with the tail, form a tripod for resting on. In flight the tail acts as a balancer, moving up and down at each leap without touching the ground.

Unlike macropod, *kangaroo* is an Aboriginal word. It was Captain James Cook who first put it into written English. On his first visit to Australia in 1770 he wrote in his journal: '. . . the animal I have mentioned before is called by the natives Kangooroo or Kanguru.'

Cook was referring to the Aboriginal people of the Endeavour River in far north Queensland where Cooktown now stands. But as kangaroos are found all over Australia and as there were more than 200 separate languages or dialects spoken then, kangaroos actually had as many names. Names like *buru*, *marlu*, *yonka*, *matyumpa*, *kulipila*, *bawurra*, now awkwardly transcribed in an alien alphabet for foreign tongues to fumble over.

Wallaby, too, is an Aboriginal word we took over early, writing it first as wal-li-bah or wo-la-ba. It seems to have been used by the Aboriginal people of Port Jackson the way we use it now, as a general term for the smaller macropods.

The first Europeans to record Australia's extraordinary marsupials were Dutch seafarers exploring the West Australian coast in the 17th century. They described two of the smaller macropods, the Tammar Wallaby and the Quokka.

Like the British who came later to colonise, the Dutchmen likened the new animals to the more familiar ones known from Europe, such as hares, monkeys, cats and so on. Later in the same century the bold English buccaneer William Dampier described a Banded Hare-wallaby as a 'sort of racoon' with very short forelegs.

Cook described the kangaroo he saw at the Endeavour River—actually a wallaby—as being like a greyhound except that it 'jumped like a Hare or a dear *[sic]*.' Joseph Banks, same time, same place, also referred to a greyhound, perhaps following Cook's lead, but he added, rather strangely: 'What I liken him to I could not tell, nothing certainly that I have seen at all resembles him.'

I suppose an Aboriginal person seeing a hare for the first time might remark on its kangaroo-like ears, and consider a greyhound a pretty odd-looking version of a wallaby.

There seem to have been no words of praise for kangaroos and wallabies from those first observers. To them a kangaroo might as well have strayed here from another planet. In its strange configuration they saw distortion. In its unpredictability they saw danger. They tried to calm their fears by looking for the familiar in the unfa-

GLEN CARRUTHERS

The Yellow-footed Rock Wallaby (Petrogale xanthopus) *in its typical habitat.*

JIM FRAZIER

ABOVE: *The female Red Kangaroo* (Megaleia rufa), *seen here with joey, is often called the Blue Flyer.*

BELOW: *Red Kangaroos are at home in the bluebush country of western New South Wales.*

miliar. Or they shot first and examined afterwards—or feasted.

Today, for us, the 'normalness' of these animals is established. We delight in the charm and elegance of wallabies, admire the proud dignity of a big kangaroo.

Less familiar are the little Bettongs and Potoroos (so-called rat kangaroos) with their unusual nesting habits. One species, the nocturnal Boodie, nests in a burrow; others make grass nests, carrying the grass curled in their tails! Its hard to believe some of these relatives of the great kangaroos are no larger than a rabbit.

The biggest of all the macropods is the male Red or Plains Kangaroo, *Megaleia rufa* ('big red'). His colour blends with the red outback earth. On the plains of bluebush and saltbush his blue-grey mate, called the Blue Flyer, blends with the foliage, though in some parts of their range she, too, has a red coat. The so-called Grey Kangaroo is more brown than grey.

As dusk falls I drive along the unfenced outback road with jumpy hands and fidgety feet. It's now, when visibility is at its worst, that the kangaroos move from their daytime resting places to feed. Roads mean nothing to them. Like cats, they lack the smallest degree of traffic sense, crossing roads without a glance to left or right. They are quite unable to judge the cheetah-like speed of this new

JIM FRAZIER

JIM FRAZIER

motorised predator across their traditionally peaceful grazing grounds.

During the hot outback summer the kangaroos spend most of the day lying up in whatever shade they can find. At cooler times of the year they bask together on the warm earth of a clearing or track. Roused by a vehicle, a kangaroo will take flight and may for some time bound ahead of it along the track.

A kangaroo fleeing fast from danger keeps its head down and its front feet together. If it's a female without a joey in pouch, you can see her small, black front paws from behind, visible between the powerful hind legs each time she

leaps and lifts her tail. Soon she'll swerve off the road to safety, but she may run parallel to her 'pursuer' for some distance.

The Blue Flyer is smaller than her mate but she can go like the wind. A full-grown male can clear eight metres with a single bound. If you stop your vehicle after putting a family party to flight, they may all stop running and turn to stare back at you over the bushes. The male stands like a giant, an impressive sight, shoulders back, enormous ears twitching, balanced for fight or flight on his hind feet and the base of his tail.

These kangaroos are not aggressive without good cause. When they have cause, watch out! Because make no mistake about it, a mature male 'Big Red' standing tall against an enemy is something to be reckoned with. The great leg muscles that send him soaring can power a slashing attack from the hind feet that will rip an assailant wide open.

On the whole, though, kangaroos and wallabies are an amiable lot. There's nothing quite so peaceful as a party of kangaroos dozing in the shade. There's nothing quite so comfortably maternal as a sleepy-eyed doe wallaby with joey in pouch, bottom weighted like a stone at rest, like a teardrop stopped, part of the old, slow, quiet Australian earth.

Even a well-grown joey will return to its mother's pouch for safe keeping.

2 Ants, lock up your children! There's a myrmecophage in the house. . .

Australian naturalist F. P. Dodd recorded just such an astonishing situation. The caterpillar he wrote about is the larva of a tropical butterfly, *Liphyrs brassolis*. Its victims and unwilling hosts are the Weaver Ants of northern Australia called *Oecophylla smaragdina*.

The ants themselves are worthy of attention with their attractive green and gold colour, and their scientific name is particularly apt. It translates as 'little emerald that lives in a leaf house'. Weaver Ants make their nests in trees, using their own larvae as living shuttles to weave the leaves together. Because their silk is used for this purpose, the larvae of Weaver Ants pupate without being able to make silk cocoons.

A Weaver Ant colony may comprise a number of leaf nests spread over a large area. A single nest may even house a thousand or more ants, larvae and pupae. But sometimes, as well as the ants, such a nest will harbour the ants' unusual enemy,

ABOVE: *When Weaver Ants* (Oecophylla smaragdina) *move house, an unwanted guest goes with them.*

Who'd want to be a caterpillar, take-away sausage snack, flavour of the month for 12 months of the year?

Caterpillars seem born to be victims in a world of hungry predators. They spend their lives turning indigestible greenstuff into bite-sized protein snacks for birds and lizards. Thin-skinned, succulent, they're a pushover for the ant hordes that macerate them with their mandibles and feed them to their nursery broods. It seems they survive only by sheer force of numbers.

Can you imagine, then, a predatory caterpillar, an aggressive carnivore, living in the very nurseries of its deadly enemies and making a daily meal of their children? Whoever heard of such a complete turn-around in the roles of prey and predator— a caterpillar that eats ants!

Well the scientific world did, back in 1902, when

RIGHT: *The hungry caterpillar* (Liphyra brassolis) *moves towards a clump of ant larvae.*

the strangely shaped, armour-plated butterfly caterpillar with a mission of mayhem.

The ants can't do a thing about their uninvited house guest. They have no stings; they attack by piercing with their mandibles and squirting formic acid into the wound. But their mandibles are unable to pierce the leathery carapace that protects the caterpillar.

So the caterpillar feeds and grows unharmed in the ants' nest and eventually pupates there, still inside its larval skin. When the butterfly emerges it climbs out onto the surface of the nest. There it rests for perhaps a whole day and night while its crumpled wings expand.

So now do the ants attack their unwanted guest in its newly vulnerable state, divested of its protective clothing? Well, they try to, but without much success, for several good reasons.

Liphyra brassolis is a member of the Lycaenidae, the family of small butterflies noted for the jewel-like colours of their wings—all shades of blue and purple—that give them the common name of Blues. This one is different. It's larger than usual and the wings are brownish-orange marked with black. When it first emerges it looks very strange indeed. Its wings and legs as well as the front part of its body are covered with a dense coat of loose white scales. And its rotund abdomen is matted with wiry grey hairs.

ABOVE: *The caterpillar drags the ant larvae under its carapace and proceeds to devour them.*

LEFT: *The pupating caterpillar fastens itself down with silk. Its leathery skin will form a shelter, or* puparium.

When the ants attack the newly emerged butterfly, two things can happen. The white scales slip off, carrying the ants with them, or the grey hairs tangle in the ants' mandibles, distracting and upsetting them. So the attack weakens and dies and the butterfly eventually flies away unharmed.

Some years ago my partner, wildlife cameraman Jim Frazier, filmed the emergence of one of the butterflies, but this footage lay unused in our library. Later, for film about Australia butterflies, we wanted to include the story of *Liphyra brassolis* as a major sequence. But a vital part of the story had not only to be filmed—it had to be confirmed.

You see, Dodd had never *seen* the caterpillars eating the ant larvae. A carefully observant natu-

The newly emerged butterfly is protected from the attacking ants by a covering of slippery scales.

ralist, he had nevertheless based his conjecture on circumstantial evidence alone. Over the more than eighty ensuing years it had never been proven. So Jim and I would have to find out the facts for ourselves. That's why for several months my house became home to three of the renegade caterpillars and a nestful of Weaver Ants.

While Weaver Ants are easy to find, *Liphyra brassolis* is not. We co-opted a Queensland naturalist friend, John Young, to help us. It took him two months of nest-checking to find three caterpillars, and you could say he risked his skin doing it. The disturbed ants, resentful at having their nests pulled apart, squirted enough formic acid to raise painful blisters all over his hands and arms.

As a food supply for the caterpillars, an entire nest of ants was transferred without too many casualties to a potted lemon tree in my sitting room. We kept the three caterpillars on my kitchen bench, in separate covered dishes of transparent plastic, raised high on a sheet of glass.

With each caterpillar we placed a clump of ant larvae together with attendant workers. A desk lamp provided tropical warmth. My magnifying make-up mirror gave a good view of the underside of the dishes and their inhabitants.

Days passed as we watched and waited. Weeks passed. The caterpillars did nothing but trundle about their prisons like clockwork toys. Sometimes the ants piled their charges and them-

selves onto the dish-shaped carapace and went trundling around with the caterpillar. If Dodd was right it was, after all, the safest place to be.

But *was* Dodd right? We'd spent a lot of time and seen no action. Perhaps, after all, the caterpillars were vegetarians. Or perhaps they fed on the excretions of the scale insects these ants often keep in their nests.

Failing anything better, Jim decided one day to film some close-ups and took one of the containers into the studio. I heard the camera whirring, then suddenly Jim called 'Come quickly, it's happening !'. I did and it was.

Clearly visible in the mirror I saw the caterpillar reach forward and drag a clump of ant larvae under its carapace, out of reach of the attendant ants. One by one the larvae disappeared between the caterpillar's mouthparts. Jim kept his finger on the button, too excited to stop. I raced for my own camera to take some photographs before it was all over.

But now the other caterpillars in the kitchen had also started eating the larvae in their containers. Our cameras ran hot. Several weeks later three well-fed caterpillars pupated.

Now Dodd in his Heaven can rejoice that his educated guess, so long a matter for controversy, has turned out to be correct. The evidence is in the can and irrefutable.

There are still problems to be solved. How do the newly hatched caterpillars find their way to the ants' nest?

Whatever mysteries remain, the story of the Weaver Ants and the turncoat butterfly stands as one of the most bizarre to come out of the ever-astonishing world of insects.

ABOVE: *The puparium after the butterfly's emergence. Note the body 'fur' left behind.*

TOP: *The butterfly at rest after its escape from the ants' nest.*

3

Fraser Island–first impressions

Crested terns (Sterna bergii) *at Seventy Five Mile Beach, Fraser Island.*

ABOVE: *A freshwater lake perched on a bed of pure sand high above the island's saltwater table.*

Magical sand of childhood memories. Twice-daily bearer of gifts from the sea. Home of crab and cockle, breakfast table of foraging bird. Stuff of sandcastles.

Fraser Island, off Queensland's south coast, is itself a giant sandcastle built by wind, sea and time and not likely to be washed away by the next wave. More danger perhaps from the rising tide of humanity, washing against it. Because Fraser Island, the biggest sand island in the world, has something for everyone and sooner or later everyone comes.

Those who come can be separately grouped by the direction and focus of their gaze.

Beach fishermen look outward, turning their backs on all but the curve of rod and line, the glint of living silver in the surf.

Conservationists look inward at the beleaguered rainforests, assessing opposition, planning strategies.

Logging men look upward, measuring height and girth, cost and profit—and over their shoulders for conservationists who dog their footsteps through the threatened forest.

Adventurers riding the island's cross-cut tracks can't look too far in front of their bouncing vehicles most of the time.

Scientists look backwards to the island's beginnings and deeply into everything, measuring and sampling, comparing and extrapolating.

Me, I look downward, at the sand that holds the island's secrets.

I am part of a film crew, on the island to record its beauty, explain its origins and sound a warning for its future in the light of logging and sand-mining. I am also a naturalist and photogrpaher, fascinated by the creatures so amazingly adapted to life on an island of sand.

Most people come here in the summer to camp, swim, fish , explore, but the big crowds come for the main fishing season—the 'tailer run' between July and October, when the big schools of tailer gather offshore for spawning.

OPPOSITE: *Spiky shadows of Pandanus trees pattern the sand beds of creeks along the island's east coast.*

LEFT: *Wind makes ripples on one of the moving dunes that form Fraser Island's uplands.*

Though some are distinctly amber-coloured, Fraser Island's lakes are considered to be amongst the purest in the world.

The fishermen arrive then by the thousands, crossing by vehicle ferry in their hired four-wheel drives. Four-wheel drive it has to be, for the beaches and the rough, sandy tracks of the hinterland take the place of roads. And if you're wise you don't bring your own vehicle to pick up a dose of rust in this domain of salt and sand or to risk getting bogged in the wet sand.

At low tide, then, along the vast 90-mile beach fronting the spawning grounds, the fishermen race their vehicles to establish their territories. In scattered groups they stand knee-deep and shoulder to shoulder in the surf, rods rampant, walled off from the rest of the island by parked vehicles and singleness of purpose.

There's beauty, though, in the brightly coloured fishing gear, in the diminishing perspective of rods etched against the sky and vanishing far up the beach into the sea mist. The Brahminy Kites enjoy it too, watching from pandanus look-outs along the foredunes, ever-present, ever-vigilant.

Handsome in chestnut and white plumage, fierce-eyed, the kites soar on lazy surveillance flights up and down the beach. They don't do much fishing at this time of the year—why bother when fish heads and offal lie high and dry for the snatching? An ever-renewable resource? Well, the kites may be forgiven for this short-term view . . .

Nothing happens on Fraser Island that isn't ultimately dependent on sand. And, wonderfully, instead of the stunted growth and paucity of wildlife you'd expect, this 'sandbank' is the unlikely matrix for a flourishing and complex ecology.

The giant trees of the hinterland—hoop pine and kauri, blackbutt and turpentine and others—are rooted in sand. Sand nourishes the cool, mossy forests of treefern, palm and vine; the flowery heathlands; the swamps of grasstree and banksia.

Here too grows Australia's largest orchid, *Phaius tankervilliae*, a swamp dweller with pleated leaves and tall-stemmed pink and brown flowers.

Sand beds the clear, shallow creeks and, unbelievably, holds more than 40 freshwater lakes perched high and safe above the saltwater table.

Sand forms the high dunes that move in slow, slow motion inland from the south-east, killing old forests as they go, creating new ones behind them.

Nothing much happens on Fraser that isn't recorded in the sand. On the tracks of the hinterland, snake, monitor and Frillneck Lizard write their names in passing. On the dunes, earwig and centipede leave tracks like miniature trail-bikes. Everywhere the island is pockmarked with the nest-holes of ants and the pit-traps of their enemies the ant-lions and tiger beetles.

On the beach, stories are scribbled twice daily between tides by crab and worm, bird and mammal, as they have been since the island began. Only more often than not, the ancient patterns are overlaid these days by an intaglio of tyre tracks, the spoor of 20th-century man.

But now most of the fishermen have gone, the summer invasion hasn't started and my colleagues and I have come in the lull to film some of the island's plants and animals.

Yesterday we saw Dingo tracks on one of the high dunes. This morning my colleagues have climbed the steep dune face with cameras to find and film the wild dogs and follow other tracks

ABOVE: *A Frillneck Lizard* (Chlamydosaurus kingii), *caught in the open, puts on a defensive display.*

they may see there. I've come down to the beach to follow a different lead. Signs of the night's activities are still legible in the sand. Dingoes have been down here too . . . clearly incised paw-marks show where a pair of them traversed the damp sand together. But the softer sand above the highest tides has been scuffed by something much bigger than a Dingo.

At night, campers behind the beaches are likely to hear heavy breathing outside their tents, or the muffled beat of galloping hooves. Wild horses—Brumbies—wander the shores in family groups feeding on the sparse, salt-encrusted sand-spattered dune grasses. The Brumbies look healthy enough, though it's said they die young from a surfeit of sand.

If you walk on the beach at night, one of the Brumby stallions may come down from the dark dunes to check you out, looming ghostly in the sea mist. But he won't hurt you. The stallions are only aggressive when they're vying for favours from the placid mares. I've heard that when two stallions fight, the stronger may drive the weaker out into the surf, and there he must wait, an unwilling seahorse, until the coast clears.

Brumbies roam the island's beaches, feeding on dune grasses and drinking from fresh water runnels.

ABOVE: *The mating of Crested Terns is graceful and delicate—and astonishingly brief.*

RIGHT: *A Pied Oystercatcher chick* (Haematopus ostralegus), *lured by its parents to safety in a cliff crevice.*

There are no Brumbies in sight now. No people. No roaring four-wheel drives. I'm alone on the beach. But how can I say that? The beach is crowded. Not with dilettante fishermen, but with those who owe life itself to the tides.

Along the beach a flock of Crested Terns stand around the wet sand on their short legs, facing the wind. They're not doing much. Except that every now and then a male bird leaps on the back of a female. For minutes at a time he balances there, wings folded or stretched high above him. He looks down at her and she looks up at him and with the utmost grace and delicacy, and astonishing brevity, they mate. No sooner does one couple part company than another goes into the act and so it goes on, tern and tern about.

We filmed on the beach yesterday, risking the Landcruiser in the wet sand to record the private lives of these terns and the expertise of a pair of oystercatchers hunting for pippies nearby. The shore birds are used to vehicles—just Brumbies with funny legs, they probably think.

Today the two oystercatchers are here again, flirting with the reluctant waves of the ebb tide. On foot I can't get as close as we could in the vehicle. Close enough, though, to see one of the birds do a bit of a pirouette on its stocky red legs. The red beak stabs down through the water into the sand

once, twice, again . . . out it comes with an impaled pippy, a bivalve that forages under the sand at the water's edge. The pippy, snapping shut too late, has closed as it was meant to on the bird's bill, to be jerked out of the sand and prised open.

The bird runs clear of the waves, drops the pippy, goes back to find another. Oystercatchers often stockpile their catches. But how do they locate their prey in the first place? Perhaps with their partly webbed feet, rather the way fishermen find the same pippies for bait by shuffling on the spot. Or perhaps those round scarlet eyes see some minute displacement of sand where the pippy lies buried.

Up on the fore-dunes these birds have a chick. They leave it hidden in a bower of gold and crimson dune-flowers while they forage. Parents of older chicks take them down to the water every evening to teach them their trade; in the fishing season it must be like crossing an eight-lane highway on a holiday weekend.

Yesterday I went looking for this baby bird to photograph it. The parents came running and flying, beep-beeping like police whistles. They tried the broken wing trick then, to lure me away. Ah, but I know about broken-wing tricks. Clever me, searching in the *opposite* direction . . . and still searching, hot and bothered, an hour later.

After I'd cooled off in the sea I heard the birds calling high on the steep sand cliff. Between them a fluffy novice mountaineer struggled upwards, fell on its beak, climbed, slid, climbed again, obedient to the parental summons.

I looked away and lost the chick entirely. I burned my bare feet climbing the hot sand and wondered how the chick's feet coped. Luck led me to the crevice where the parents had stowed their one-and-only. While they dived and screamed I took my photograph, bearing for their fear and anxiety the faint feeling of guilt that goes with being a wildlife photographer.

It's different with crabs. I don't think their little hearts beat too wildly as they scuttle for cover. They're just cautious. And it's crabs I'm concerned with today. Soon we'll be moving camp to the hinterland to film rainforest and lakes. Before we leave the beach I've promised myself one thing: to find out what makes Sand-Bubbler Crabs tick (or bubble) and to photograph them doing it.

RIGHT, TOP TO BOTTOM: *The oystercatcher pulls a shellfish from the sand at the tide's edge, prises the hinged shell apart with its beak and eats the contents.*

Feeding on detritus in the sand, the Sandbubbler Crab (Scopimera inflata) *patterns the beach with its leftovers.*

Crabs dominate life on the beach here as elsewhere. The sand I'm standing on is the roof of their hidden city. Most of the crabs are nocturnal. Others set their clocks by the moon and work on the falling tide. Being vulnerable in daylight, these are mostly small and cryptically coloured. Among them are the Sand-Bubblers.

Sand-Bubbler Crabs are no bigger than my thumbnail, but, massed together, they can change the texture of an entire beach in a few hours. With their intricate patterns of sand pellets they can claim to be the craftsmen of the crab world, sculptors in sand. Yet all they do is feed.

The tide's going out quickly now and I can see the first patterns appearing: lines of 'bubbles' alongside shallow grooves that radiate from tiny burrow exits. But when I approach, the crabs themselves vanish. Their stalked eyes see me coming from six metres away. Some, braver or hungrier than the rest, reappear quickly so I choose one of these to watch and photograph.

Lying on the sand, camera ready, I wait. The crab pops up and it waits. I hold my breath. The crab flicks its eyestalks from horizontal to vertical. The eyestalks are hinged like portable TV antennas. They're lowered when the crab burrows, so it won't get sand in its eyes. Now its eyes need to be high up to watch for danger as the accumulating sand bubbles cut off the view.

Down at crab level I can see that a Sand-Bubbler isn't the plain jane I thought it. Sand-coloured, yes, but with touches of blue and scarlet. Even a

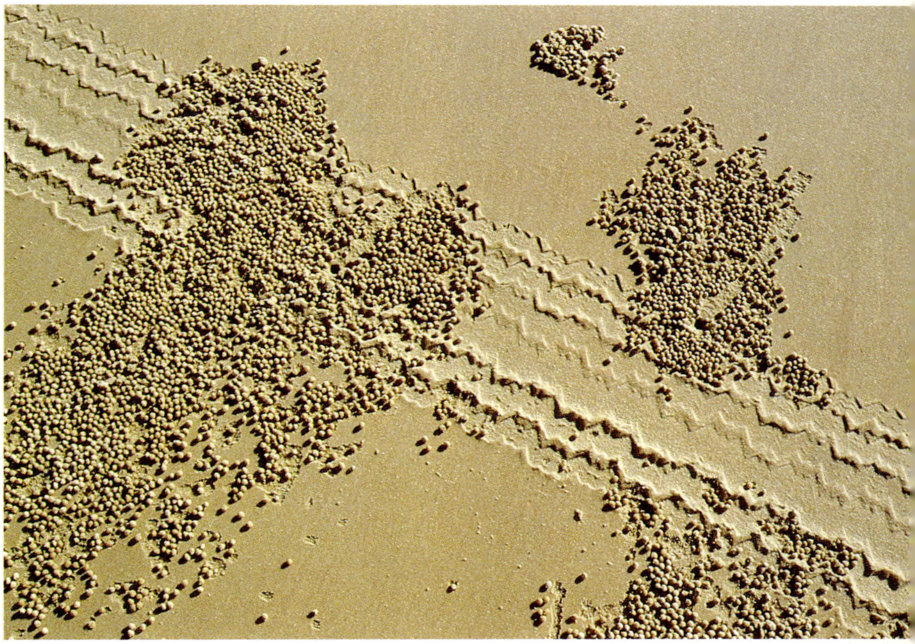

At low tide, crab and man leave evidence of their activities on the sand of Fraser Island's beaches.

flush of royal purple on the pincers. Nothing to be ashamed of, Jane Crab, not at all. Anyhow an artist, a modeller in sand, doesn't need to be beautiful.

I keep still enough for the crab to start feeding, stuffing sand into its mouth with tiny pincers. Eating sand? Well, that's what it looks like. But at the same rate that the sand goes into the crab's mouth it comes swelling out again like bubble-gum.

I risk a photograph. The crab dives underground, dropping its bubble-gum, clearly seen now as a solid ball. I wait again. My skin prickles from the damp sand, my elbows are raw. March flies have homed in on me and I can't slap at them. It's march flies that keep people away from this island from November well into summer.

Up comes the crab. Down and up go the little pincers, one-two, one-two, picking up sand. This time the crab wipes the fully formed sandball neatly off its face and to one side, kicking it backwards through its legs like a football in a Rugby scrum. It moves forward a little and starts making another one.

For every sandball the crab makes and discards, it has travelled a little further from its retreat hole. As it progresses it leaves a continuous scrape mark or furrow and a line of 'bubbles' behind it. What happens is that the complex mouthparts of the crabs act as a sorting machine. The edible organic bits and pieces washed up on the beach by the last tide are sorted out from the sand grains and swallowed, and the cleaned sand is returned to the beach. Every so often the crab returns to the hub and starts a new feeding track so that a radiating pattern is built up around the burrow.

By low tide, when all the Sand-Bubblers have been out for a meal, the entire beach has been sieved clean. The combined effort of thousands of crabs has produced an abstract sculpture in low relief, enhanced by the long shadows from the late afternoon sun.

I'm enchanted and I'm itchy from sand and march flies and my back is burning and one foot has gone to sleep and I've run out of film.

And I'm ready now to tackle the rest of Fraser Island, the magic sandcastle.

4 Give a frog a bad name. . .

A sprinkling of emerald green dots—Peron's Tree Frog (Litoria peroni).

What's in a name? Well, we don't need Shakespeare to tell us that a rose would smell as sweet if we called it a pumpernickel. And a frog doesn't turn into a handsome prince just because someone gives it a high-sounding name such as *Taudactylus acutirostris* or *Limnodynastes peroni*.

Actually the last of those names comes close to being princely, or at least aristocratic. The first part of it means 'lord of the marshes'. The second part of the frog's name belonged originally to a 19th-century French naturalist, a M. Francois Peron.

The early enthusiasm to check out the fauna and flora of this new continent was shared by many Europeans, including the French. It's not commonly known what a large part they played in the naming of Australian animals. Most of this was done in Paris from specimens sent over by collectors in Australia.

In 1802 a French ship under Captain Nicolas Baudin arrived in Australia on a voyage of discovery and adventure. On board were two people whose names are very familiar now to anyone interested in our natural history.

Francois Peron was a naturalist. His friend and travelling companion Charles-Alexandre Le Sueur was a scientific artist. Between them these two young men sent back to Paris more than 100,000 zoological specimens. There they were described by others and given their formal scientific names.

Some of these double-barrelled names commemorate in their second part one or other of the two young Frenchmen, the original collectors. For instance, among reptiles and amphibians there are two frogs, a lizard and a sea-snake called *peroni*. Another frog and two lizards are called *lesueuri*. There's no confusion because for each animal the first part of the name, the genus name, is different.

Limnodynastes peroni, *Litoria peroni* and *Litoria lesueuri* were by no means the first Australian frogs to be described and named. That honour belongs to the Green Tree Frog, the big one with the Mona Lisa smile, often found around human dwellings. This frog was described by an Englishman. The French might have done better, because this bright green frog was named *Litoria caerulea* (*blue* tree frog). They say it was the frog that was pickled . . . Jokes aside, though, it really was pickled, in alcohol. That's what turned it blue.

OPPOSITE TOP: *Only blue when pickled—Green Tree Frog* (Litoria caerulea).

OPPOSITE BOTTOM: *As pretty as its name—Green and Golden Bell Frog* (Litoria aurea).

Glamour-puss of the tree tops—Red-eyed Tree Frog (Litoria chloris).

Apart from the ones named by scientists after other scientists, perhaps in the hope of having the favour returned, most of our frog names are descriptive. Like *Taudactylus acutirostris*, which as anyone can see means 'sharp-nosed frog with T-shaped toes'. So why not say so in plain English, I hear you complain. But the English version has six more letters, three more spaces and two hyphens. And how splendidly those syllables roll off the tongue after a bit of practice! Not that the tiny *Taudactylus* cares a fig, tinkling away on its rain-forest mountain.

Well, as I said, what's in a name? The real answer is: clarity in communication. Without an agreed scientific name in an agreed language how could scientists in different countries speaking different languages know exactly which frog was being referred to? How could we know what they're talking about? Common names are for people who just like frogs and are prepared to put up with the confusion when the same frog has different names in different place. Or when several frogs in different places have the same name. Trouble is, our frogs are stuck with some pretty drab common names. That is, apart from Pobblebonk, so descriptive of the sounds made by a pondful of those frogs. And perhaps Corroboree Frog named for its aboriginal type pattern.

But it's a pity about all those plain-jane names like Brown Frog, Green Tree Frog and so on. Because however ugly and long-winded the scientific names might seem (and I don't agree for one minute that they are), the owners of the names are mostly colourful and have charming characters. Close up, Peron's Tree Frog has its plain brown back sprinkled with emerald green dots. The Green and Golden Bell Frog (not a bad name) wears a pattern of chartreuse and copper with a touch of aquamarine on the flanks. The splash of lilac on the thigh of the Slender Tree Frog from tropical Queensland is like a brilliant artistic afterthought, and the weepy-sounding Red-eyed Tree Frog is a clear-eyed beauty.

Very often a frog you thought dull will give you a surprise when it leaps away. Unexpected patches of bright colour revealed in armpit and groin and flank are sometimes called 'flash colours'. You get flashers among the grasshoppers, too. It's a technique for confusing the eye of a predator. As the potential meal leaps away the colours flash out like a travelling neon sign, then when it comes to rest they disappear. The enemy's eye travels onwards, seeking the telltale identifying colours, and the game is lost to the prey. In a paradoxical sort of way you could say that a frog using flash colours to evade an enemy is acting as its own decoy.

Getting back to the history of herpetology in Australia, how appropriate for our frogs—for any frogs—to be discovered and named by Frenchmen! No, I am not being derogatory, because the English slang term 'frog' or 'froggy' for a Frenchman actually has quite honourable origins. It is thought to be derived from the frogs or toads shown on the shield of the City of Paris.

The dates may be way out, but I wonder if Charles-Alexandre Le Sueur, painter of animals and discoverer of down-under frogs, had any hand in the design of that shield.

A touch of lilac on the thighs—Slender Tree Frog (Litoria gracilenta).

LEFT: *Hidden colours flash out like a neon sign—Whistling Tree Frog* (Litoria verreauxii).

5 Cupmoth caterpillars may give you the brush-off

Woolly-bear and hairy mary; cutworm and cabbage-worm; inchworm, looper, saddleback and Chinese Junk. All common names of some of our caterpillars—the wingless, creeping, crawling, munching young stages of moths and butterflies.

A cupmoth caterpillar (Doratifera vulnerans) *raises stinging spines as a defence against predators.*

Usually we give common names only to insects that affect our lives in some way. Butterflies that delight us. Bugs that bite us. The beauty of caterpillars, compared with their sometimes dull adult forms, is often overlooked. They're not all plain green sausages, and most are harmless, though a few make a nuisance of themselves.

Woolly or furry caterpillars, often stunningly beautiful, admittedly make us itch. Cutworms and cabbage-worms admittedly eat our crops. But most garden and agricultural pests are introduced from elsewhere. Australian caterpillars usually prefer native food-plants.

Loopers or inchworms—same thing—are mostly harmless enough. They just attract attention by the delightful way they move, alternately looping and stretching their long bodies, with legs fore and aft but none in the middle. The official name for the looper family of moths is Geometridae, which means 'earth-measurers'.

Chinese Junks or saddlebacks are caterpillars of the cupmoth group. They get their name from

ABOVE: *Looking remarkably like gumnuts, the cocoons of cupmoth caterpillars are easily overlooked.*

their exotic shape and the way they glide along like ships at sea. Less romantically, they're sometimes called slug caterpillars. In a way that's an apt name because they look rather like an air-breathing version of those glamorous and colourful slugs of the sea, the nudibranchs. The colours are a warning. Cupmoth caterpillars can sting.

Encounters with unknown stingers and biters must have been alarming to the early explorers. Who could tell what was lethal and what was not? Some stinging caterpillars were referred to in Joseph Banks' journal during Cook's voyage up the Queensland coast in 1770. The ship *Endeavour* called into Bustard Bay, and a number of landings were made. Among the natural history sights (and sensations) recorded by Banks was the following:

. . . a small kind of Caterpillar, green and beset with many hairs; these sat upon the leaves many together ranged by the side of each other like soldiers drawn up, 20 or 30 perhaps on one leaf; if these wrathful militia were touched but ever so gently they did not fail to make the person offending them sensible of their anger, every hair in them stinging much as nettles do. . .

Some think these were some kind of cupmoth caterpillar. Certainly when the caterpillars are young and quite small they feed together side by side, and they do look rather like platoons of soldiers. But the reference to 'every hair' doesn't seem to fit the cupmoths' array of stings.

Most people with gumtrees around their place know about cupmoth caterpillars. They wander about our garden plants looking lost, blown from a tree by the wind or dropped by a bird that got

Cupmoth cocoons have a lid that pops open when the moth is ready to emerge.

ABOVE RIGHT: *The caterpillars have hatched under their fluffy blanket, eaten their first meal and gone.*

BELOW: *Baby cupmoth caterpillars feed only on the upper surface of a leaf.*

more than it bargained for. The caterpillars won't eat garden plants, so they're likely to starve.

You've heard of a cat on a hot tin roof. What about a caterpillar on a hot car roof? It happens to cupmoth caterpillars all the time in summer. They drop off trees on top of cars parked in the sun and they usually frizzle up before you can rescue them.

Any ordinary caterpillar would sprint off the hot metal the way we'd sprint off hot sand. Cupmoth caterpillars can't sprint. You'll see why if you put one on your window and watch its underside. It just ripples along on its stomach like a snail. The three front pairs of legs are just tiny claws. And all those extra legs most caterpillars have are missing. When it comes to hot cars, it's the looper caterpillars that get the best deal.

In the garden you'll most often feel a cupmoth caterpillar before you see it. Brush your arm against a bush you thought you could trust, and suddenly it turns on you. You leap away with a burning, prick-

ling sensation on your skin. But don't glare at the bush. Look closely and you'll see the real culprit sitting on a leaf, still armed and ready to fend you off.

Don't be too hard on the caterpillar; its armament is for defence only. The eight sets of stinging spines, four at each end and sharp as needles, usually lie flat and harmless, like bundles of toothpicks. Touch the caterpillar's smooth skin anywhere and the spines rise instantly to become eight prickly pompoms. Whichever side you touch, the caterpillar will lean towards your finger, trying to reach you with all its spines at once.

As Banks suggested, when you brush up against one of these caterpillars it causes no more discomfort than if you brushed against a stinging nettle. The pain soon wears off. Then you'll feel you can ease your finger under the caterpillar, pick it some nice fresh gum leaves and take it indoors where you can watch it.

When your caterpillar stops feeding and starts gliding up and down the stems you'll know it's looking for a good place to make a cocoon—the sort of place where a gumnut would look suitable as long as you weren't a gumnut expert. Birds that eat caterpillars aren't gumnut experts.

The first thing the pupating caterpillar does is fix some silk to a stem. Then it does something rather strange. It curls up in a circle, turning its backside inside and its underside outside, so becoming more or less spherical. Now, paying out the silk from a spigot on its chin, it weaves a silk bag around itself, rotating inside the bag as it goes. The smooth, round body acts as a form for shaping the cocoon.

The finished product is a beautifully symmetrical container, pear-shaped with a built-in lid at the narrow end. All that remains is to flood the fabric with a fluid that will both darken and harden it.

If you keep the cocoon under surveillance—perhaps near the kitchen sink—you might see the lid pop up one day, and a small brown moth climb out and fly away.

ABOVE LEFT: *A cupmoth caterpillar* (Doratifera oxleyi) *'turns inside out' to make its cocoon.*

BELOW: *Most cupmoth larvae are attractively marked, like this Painted Cupmoth caterpillar* (Doratifera quadriguttata).

To the Southern Leaftail Gecko (Phyllurus platurus), *a brick house is just another rock cave to shelter in.*

OPPOSITE TOP: *Not a snake but one of a number of harmless legless lizards found in Australia* (Delma fraseri).

OPPOSITE MIDDLE: *The Wood Gecko* (Diplodactylus vittatus) *intimidates predators with a display both visual and vocal.*

6 Lizards with contact lenses and windscreen wipers

To the leaftail geckoes my house is probably just another rocky outcrop in Sydney's northern bushland. The basement is a cool, dark cave to hide in by day.

Outside at night the brick walls give rough foothold for creeping, courting, stalking and pouncing, and of course there's the added luxury of built-in insect lures, the lighted windows.

Soon after dark I watch the first of the geckoes leave the basement snout-first through a ventilator, one round, delicately veined eye checking for danger. There's no hurry, the hunting's good on a warm summer night. I do the rounds, checking all the ventilators. Some have mesh behind, put there when the house was built. Most are open between the upright bars, making exit holes for the geckoes to slip through. Sixteen is the highest count so far.

The geckoes take up their ambush stations low down on the walls, flat against the off-white bricks, claws spread to grip and hold. Which end is which? Front and hind limbs are mirror-imaged with elbows and knees angled towards each other. Although the broad tail comes to a point, and the snout is rounded, at a quick glance the gecko could be facing either way.

Motionless against the wall, the geckoes appear to be not so much living animals as stylised artefacts, the creation of some ancient, exotic civilsation. Central American, perhaps: the shape goes well with spiky names like Aztec, axolotl and Quetzalcoatl. And what about *gecko* itself? It seems to belong on that list, especially as spelled in the family name Gekkonidae.

But no, gecko comes from the Malayan word transcribed into our alphabet as gekoq. It's based on the sound made by one of the common house geckoes of South-East Asia. Among reptiles only the geckoes seem to have become inquilines in human houses, sharing man's shelter though not his food. Most tropical countries, including northern Australia, have their domestic geckoes. They creep out from behind pictures and curtains and lavatory cisterns after dark; pale, black-eyed little extroverts that fight and mate and mess up the walls and ceilings quite openly. And noisily.

Vocalisation in lizards is rare. Only geckoes and their close relatives, the legless or flap-footed lizards, can produce sounds from their mouths. Have you ever heard a Bluetongue bleating or a goanna groaning, or a Frillneck fulminating? I didn't choose that last word solely for the sake of alliteration; Frillnecks are aggressive little animals and I'm sure they would fulminate thunderously if they could.

No, a hissing and a rustling is about all you ever hear from those little lizards. Or at most the clash of jaws on armour-plating when two male Shinglebacks come to grips in the outback, or try to. But many outback geckoes open their mouths and bark when they're attacked. Disturb a Knobtail and it will swell up and 'yap' aggressively while it does press-ups or makes threatening little runs at you. All bluff, of course. Its bite isn't worth an ouch.

If bluff and bluster don't work, a gecko has a second line of defence. Grabbed from the rear, it

ABOVE: *Like all geckoes, a Leaftail can escape an enemy by dropping its tail. A new tail grows in its place.*

can voluntarily sever its tail, drop it, and make a quick getaway. Geckoes share this remarkable ability with the skinks and the scalyfoot lizards. A new tail grows again after a while but it never looks quite the same. Two of the leaftails around my house at present have shiny, dark brown tails, no longer leaflike but more like some kind of seed-pod. Somebody's car is probably to blame. Not mine.

But if you do come across a leaftail or any other gecko with its skin hanging around it in shreds, don't imagine it's been attacked. This happens regularly and it's just the normal way for a gecko to shed its outgrown skin.

The leaftails don't come indoors much. Not that the hunting isn't good—they just don't have the feet for it. Many geckoes have specialised feet with expanded, disc-like toes that give them traction on smooth surfaces. That's how tropical house geckoes cling to window panes and run upside down on ceilings. Leaftails belong to the other group of geckoes that have more bird-like feet. They depend on their claws to grip irregularities in rough rocks and tree trunks. But in my house some of the inside walls are brick, so it's not unusual for the occasional gecko to take up temporary residence.

I mentioned the scalyfoot lizards of the family Pygopodidae, sometimes called snake-lizards. These little reptiles have neither claws nor toepads nor even limbs, so they're confined to the

ABOVE: *Under attack, this Knobtail Gecko* (Nephrurus asper) *will put on an aggressive display that's mostly bluff.*

RIGHT: *A sorry sight, but this leaftail is simply going through one of its periodical moults.*

ground. All they have left in the way of limbs is a pair of scaly flaps just in front of the vent, vestiges of hind legs that still contain tiny bones.

As a rule the flaps on a scalyfoot are small and hard to see because they're held close to the body. But a few pygopodid species with bigger flaps stick them out when they're disturbed. It's almost as if they still had the urge to run rather than slither.

One of these lizards is the Common Scalyfoot, called *Pygopus lepidopodus*. The second word is just Greek for scalyfoot. Scalyfoots are daytime hunters of insects. I usually find them coiled up under a rock, and when I do my first thought is: snake or lizard? Not that I'd react any differently; snakes are welcome in my garden. It's just that you need to be a little careful.

Pygopodids move like snakes and look superficially like snakes and more often than not they get themselves killed because of it. But they're strictly harmless. No Australian lizards of any kind have venom. They kill their prey by chomping on it.

There are several ways to tell a snake from a snake-lizard or scalyfoot. For one thing, no snake has an external ear, while most snake-lizards have an ear-hole just behind each eye. But the best way to find out is to watch until it puts its tongue out to get some information about you. If the tongue is narrow and forked it's a snake. A legless lizard, like a gecko, has a broad tongue, with a slight notch at the tip.

Speaking of tongues leads me to eyes. There's something both geckoes and legless lizards do have in common with snakes: they can't blink. Instead of movable eyelids they have fixed, transparent, built-in 'contact lenses' that protect their eyes against dust and dryness. These peel off with the rest of the skin when the lizard moults, leaving a replacement pair underneath.

But these lenses are themselves vulnerable to dust and to splashes of rain and no doubt they fog up at times, so it's useful to have something like a built-in windscreen wiper. For geckoes and legless lizards, the windscreen wiper is the long, straplike tongue. It's licked upwards from the corner of the mouth in slow, easy stages until one eye is reached and wiped clear. Then it does the other eye.

This is an arrangement that can only be envied by the bespectacled photographer when a raindrop or speck of dust or a sudden fogging up is about to ruin some never-to-be-repeated wildlife shot.

ABOVE: *Geckoes and legless lizards are unable to blink; instead they use their tongues to clean and moisten their eyes.*

TOP: *A legless lizard, the Common Scalyfoot* (Pygopus lepidopodus) *with its eggs.*

7 Brolgas dance the light fantastic

***Brolga*. It's an abrupt sort of name for that most elegant of dancing birds, our native crane. But *Brolga* was originally *Buralga*, a famous dancer of Aboriginal legend.**

Changed into a bird by the sorcery of a rejected suitor, Buralga or Brolga still dances in company with her friends. Her dance steps have long been woven into the ritual choreography of the Aboriginal people of northern Australia.

The dances of various kinds of cranes are mimicked in other countries, too. The Ainu people of Japan use the display of the Japanese Crane. For certain African tribes it's the Crowned Crane. Birds of the crane family are found in every part of the world except South America, and they're all accomplished dancers.

There's another remarkable thing about cranes. They've all got a tremendously long windpipe, coiled up like a French horn inside the hollow breastbone. In the Whooping Crane it's 1.5 metres, long enough to produce a pretty impressive whoop. Our Brolga, too, is no mean horn player, though the sound it makes is more often referred to as trumpeting.

The first white man to see an Brolga was probably Joseph Banks. There's only a brief mention in his journal, recording '. . . cranes. . . of which we saw several very large and some beautiful species'.

That was in north Queensland, but in Banks' time the Brolga was a common bird around the swamps of eastern as well as northern Australia. In fact they could be seen in numbers around Sydney and Melbourne and it's a matter of great regret that they haven't stayed around to establish a closer relationship. But it's said their numbers are increasing again in some of the southern areas.

Although Banks mentioned 'several species', there are actually only two kinds of crane in Australia. What we commonly call the Blue Crane or White-faced Heron is in fact a true heron. As well as the Brolga there's the closely related Sarus Crane. But the Sarus Crane wasn't recorded in

JIM FRAZIER

JIM FRAZIER

ABOVE: *The elegant Brolga* (Grus rubicundus) *patrols the swamplands of northern and eastern Australia, feeding on sedges and insects.*

RIGHT: *The Brolga is one of only two members of the crane family found in Australia.*

OPPOSITE: *Between meals, a Brolga attends diligently to its toilet.*

JIM FRAZIER

JIM FRAZIER

JIM FRAZIER

ABOVE: *Brolgas are companionable birds, not often seen on their own.*

TOP: *The Brolga uses its remarkably long, coiled windpipe to produce a sound like a trumpet.*

Australia until 1966. Before that it was regarded as a South-East Asian bird.

The two species are similar. The Sarus Crane is a little bigger than the Brolga, with more red on its face and neck. It has pinkish-red legs where the Brolga's legs are dark grey. So far Sarus Cranes have been seen only in northern Australia.

The white man's early name for the Brolga was a strange one: Native Companion. I find it as puzzling as the name Miner's Friend given to those chewy, tangy confectionery bars that were around when I was a child.

Perhaps the name has something to do with what the early settlers had learned about Aboriginal involvement with the birds' dance routine. Or perhaps it refers to the companionable way Brolgas have of dancing together. It's one of those intriguing examples of adult 'play' that seems to be done for no other purpose than plain enjoyment.

Whatever the reason, it's delightful to watch these slender and graceful birds put on a dancing display. One bird starts. Another bird catches on, then the entire flock. The urge to dance spreads through the entire flock. There's a tossing of heads and a deep bowing. Sticks are picked up from the ground and thrown into the air. High stepping leads to vertical take-off as wings lift and beat, faster and faster. Gently the bird floats down

again, to leap and pirouette with utmost grace.

Between pairs the display is part of courtship routine. Brolgas mate for life, so it could also be thought of as a reinforcement between already bonded pairs.

Dancing is part of our own courtship ethos, mostly used to bring male and female together and give them a chance to look each other over. Perhaps for us, too, it can help reinforce bonds already tied. But mostly, don't we just dance for the sheer joy of it? And who are we to deny that Brolgas might do the same?

The late poet and humorist Leon Gellert, in his little book of poems called *These Beastly Australians*, said this about Brolgas:

The Brolga or Australian crane
Holds dancing orgies on the plain;
And students of the Russian Ballet
Have crept in crowds across the mallee,
Far out beyond the desert grasses
To where the Brolga holds its classes;
And many a motif for the leg
Has emanated from an egg.
But certain purists do not care
For such techniques and prudes declare
Specific movements of the Brolga
Unacademic, if not volga.

Noted for their elaborate courtship dances, Brolgas perform outside the breeding season also.

8 Industrial espionage in the spider world?

Have you met Maggie? She's a lady worth watching. You might find her in your garden after dark, fishing for insects with a short silk thread dangling from one leg. A spider, shining white in the torchlight.

ABOVE: *She seduces male spiders with a promise of love—the Magnificent Spider* (Ordgarius magnificus).

OPPOSITE: *The trap is a simple sticky thread, the bait a complex chemical.*

Look closely at this spider. You'll see she has patterns of delicate orange and yellow on her rounded body. Her eight legs are banded in coffee and cream, her eight eyes rise on a crimson turret from her carapace. Who said spiders were ugly?

Now look closely at her fishing line. Even with its coating of sticky droplets, even with that extra big blob of glue at the bottom, it seems much too simple to catch anything, let alone a powerful flying insect.

'Maggie' weaves her silken trap after dark.

What you can't see, though you will if you watch long enough, is how the spider turns her fishing line into something more like a large butterfly net. And what you can't read is the message the spider's sending out to lure a special kind of prey—not a butterfly, but a moth. It's written in the language of scent and it purports to come not from the spider but from another moth.

The Magnificent Spider, *Ordgarius magnificus*, is a member of the orb-weaving family, like the common Garden Spider. But she's a renegade. She no longer makes an orb web because somewhere back along the evolutionary path her orb-weaving ancestors branched out to become specialists.

You might think the use of an orb web is already a specialisation compared with, say, chasing after your food on foot as a Wolf Spider does. And so it is. An orb web is a miracle of engineering and craftsmanship. But a lot of time and energy and silk go into the making of a web, and some of the insects that are caught by it could be dangerous or unsuitable. Sometimes repairs, even a whole new web, must be made during a catching period.

But Maggie's catching device takes no time at all to make. It uses very little silk. And her victims without exception are harmless and good to eat, because the Magnificent Spider is a specialist among specialists. Her forte lies not in engineering but in chemistry; not in a catch-all technique but in selective predation on male moths of a particular kind. Except by rare accident,

no other insect ends up on her baited line.

So how does she do it? Well, here's the mystery of her system made clear, and it's no less of a mystery and a marvel for that. The essence (note the word, it's very apt) of the Magnificent Spider's technique is a confidence trick. To understand how it works you need to know a little bit about moths.

Most male moths, including the ones this spider catches, track down their mates by following a scent released by the female moth.

She may be far away but a male moth can detect mere molecules of scent with his antennae.

The female of each species of moth has her own particular perfume, so there can be no mistake, no mismatching of species. Or can there?

The fact is that sometimes a male moth finds himself led up the garden path into the embrace of quite the wrong kind of female. No moth this, but a spider that has lured him *with a scent that is a facsimile of the scent of the female moth*.

Deception of a visual kind as a defence against enemies is well known, of course. A moth suddenly lifting its wings to expose two false eyes, for instance, can frighten away an attacking bird. Or a looper caterpillar can be overlooked because it looks like a twig. Visual mimicry is something we can understand because it's directed at birds, and humans and birds both depend mainly on sight. But because our sense of smell is so poor the *chemical mimicry* of the spider is too subtle for us to detect.

What we can see, though, is the sequel to the confidence trick: the arrival of a moth at the end of his disastrous flight path, and his entanglement and capture at the ambush site. And this part of the story is as remarkable as the rest.

The Magnificent Spider is one of a small, elite group of spiders, all using sex scents as bait for male moths. Two of these are Australian, while several others, distantly related, are American. Because of the way these spiders land their victims after luring them, they all go under the general name of 'bolas' spiders, but as I'll explain, the name has been falsely given.

A bolas is a hunting weapon once used by Native South Americans. Three stones attached to one another by short cords were fixed to the end of a longer cord, which was coiled like a lasso. A large animal such as a Guanaco was stalked on foot, and the bolas was flung on a low trajectory to tangle its legs and bring it down.

Maggie's 'bolas' is more like a sticky flypaper with an extra blob at the bottom. But both analogies fall down the way she uses it. Instead of stalking her prey and throwing her weapon, or waiting passively, she simply swings it round and round.

Thus the single line becomes a cone-shaped trap. Even a moth that's flying slightly off beam can't avoid it.

When my colleague Jim Frazier and I set out to film Maggie's extraordinary story for television a few years ago, we had two main problems. The first was the availability of our leading lady. Magnificent Spiders are few and far between. Moreover, wildlife film actors can never be depended on to perform on cue, and they're quite likely to walk out on you or get themselves eaten. The wise director makes sure he has several stand-ins ready on the set.

So it was with Maggie, as we soon started calling her to save breath. In the end we had to use five (or was it six?) separate Maggies to tell the complete story.

The moths posed the other main problem. For technical reasons you need extra bright light to film tiny animals like insects and spiders, and moths fly to bright lights. Maggie would only swing her line on the close approach of a victim. So how could we film that marvellous action if the moth that should have triggered it was cooking on our movie lamp instead ?

It wasn't known at that time exactly what it was that set the Magnificent Spider swinging just as the moth came within reach, but over long nights of watching and waiting we'd picked up a clue or two. When a heavy vehicle drove past, or an aircraft flew overhead, our Maggies all responded with a few half-hearted swings. So we guessed it was vibrations. The beat of the moth's wings.

As a substitute Jim tried humming, *forte basso profondo*. I joined in, *contralto*. It worked. The spider liked the feel of it so much she swung wildly, not just round and round but up and down like a yoyo. But humming at the top (or bottom) of your hum for any length of time is quite exhausting. By the time Jim was ready to film we were out of breath. So I fetched Jim's guitar and tried strumming.

The low E string did the trick. It worked perfectly and we got our film sequence. So the spider that uses chemical mimicry to delude male moths had the tables turned on her. We tricked her with *acoustic* mimicry.

Maggie didn't get her moth that night, but she got a good fishing story: 'You should have *heard* the one that got away!'

OPPOSITE TOP: *The spider whirls her weapon as a moth flies close . . .*

OPPOSITE BOTTOM: *. . .then hauls up her catch to eat at leisure.*

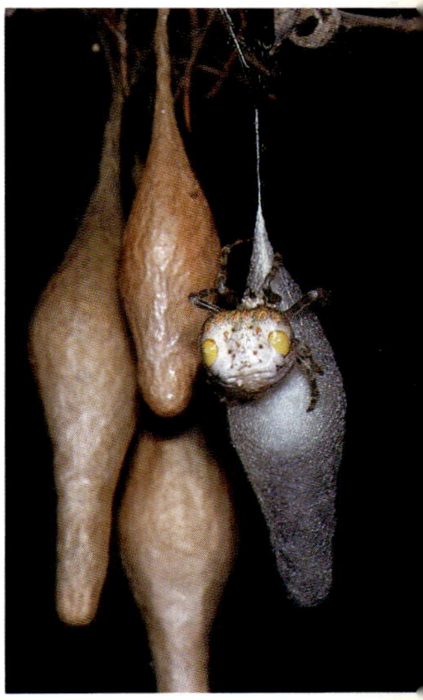

Thousands of metres of silk go into the making of these egg-sacs.

*The Blackberry (*Philomastix macleaii*) cuts a hole before pushing each egg through the leaf.*

9 Sawfly skills and strategies

Next time you walk past a blackberry bush and it makes a sound like a raspberry, don't pretend you haven't heard. Stop and listen and track down the sound—you might be surprised.

Actually it's more a short, sharp buzz than a raspberry, but it's certainly directed at you. It's a warning to keep away, coming not from the bush itself but from a rather pretty insect, orange and black and white, with shiny wings. You'll find it on a leaf resting at an angle with head down and legs clasping the leaf stalk. Wait for the next buzz and you'll see how the raised and partly opened wings vibrate to produce the sound.

This is the female of a sawfly called *Philomastix macleaii*; it has no common name so I call it simply the Blackberry Sawfly. But perhaps you don't know what sawflies are. To begin with, they're not flies at all. They're closely related to wasps, but don't let that put you off, because sawflies don't have stings. There are other differences. For instance you don't find social organisation among sawflies. They're thought to be rather like the ancestral hymenoptera, that is, the primitive forebears of all the wasps, bees, ants and present-day sawflies.

There's another major difference: sawfly larvae are active and self-supporting. This is in direct contrast with the helpless, legless, cossetted grubs of their near relatives. And they're vegetarians. I suppose you could call a bee grub a vegetarian, too, since it's fed on pollen and nectar, but the larvae of most sawflies chew leaves. In fact they're often mistaken for the caterpillars of a moth or butterfly.

But back to the blackberry bush. Just why is the sawfly sitting on top of the leaf with her legs holding the stem, and why does she refuse to budge even when you bump the bush?

Well, all the sawflies are rather sluggish compared with their volatile relatives. They even fly slowly. But this one has a special reason for staying put. The reason can be found on the underside of the leaf she's sitting on. Have a look without disturbing the sawfly. You'll see what looks like a lot of little purple jelly beans dangling there. These are the sawfly's eggs and she's there to watch over them.

But the most remarkable thing about this sawfly is not so much the protective care she gives her offspring as the way she produces them. She actually lays her eggs through the leaf from above. You'd think it would be simpler for her to go underneath and fix them in place, as many insects do. Be assured there must be a good reason; she wouldn't be doing it otherwise.

Sawflies get their name from the special egg-laying apparatus of the females. Underneath at the rear end, hidden in a groove, lies a tiny saw-like implement. It's used for cutting through or

Unlike wasp grubs, the larvae of sawflies are self-sufficient vegetarians.

into leaf tissue to make a place for the eggs.

Some sawflies hide their eggs inside the leaf tissue itself, between the upper and lower 'skins' of the leaf. Others leave the eggs exposed. Whatever the method, the operation is carried out with astonishing precision—and it's all done by touch.

Our Blackberry Sawfly goes about her egg-laying this way. Standing on the upper surface of the leaf, she makes a tiny hole in it, cutting through until you can see the tip of the saw protruding underneath. Then she squeezes the long, narrow egg itself through the opening to the underside. The egg is bigger than the hole, and it's soft, so you can see it constricting as it passes through. Down it goes, though, finally popping out to hang free under the leaf.

But before she lets go of the egg the sawfly releases a drop of sticky liquid that immediately spreads around the attachment point. The liquid sets like glue, fixing the egg in place under the leaf.

Now the insect up above withdraws her saw and moves away a short distance. Down comes another egg. And so on. When all her purple jelly beans are in place she takes up her guard position. Let me add that the eggs are not invariably purple. I've seen one female laying green eggs, while another, apparently of the same species, was laying purple ones on the same bush.

The egg-laying site could just as well be on a native raspberry bush, the sawfly's natural food. Blackberries, being closely related to raspberries but not native to Australia, are just a convenient foreign food substitute.

The female sawfly will stay around even after the eggs hatch. Then, before her translucent, fork-tailed offspring have finished their first meal, she drops to the ground and dies.

Less elegant but more familiar are the Gumtree Sawflies of several kinds whose caterpillar-like

ABOVE: *A sawfly of the gumtrees* (Perga affinis) *inserts her eggs inside the tissue of a gumleaf.*

LEFT: *The larvae are very sociable and keep closely in touch by tapping with their tails.*

A female Lewis's Sawfly (Perga) *guards her incubator —a 'blister' on a gumleaf.*

larvae feed on eucalyptus leaves. Gumtree Sawflies lay their eggs in quite a different way from Blackberry Sawflies. The thickness of a eucalypt leaf makes it possible to have a totally enclosed incubator for the eggs.

Typical of the Gumtree Sawflies is the rather ordinary brownish female of Lewis's Sawfly, *Pseudoperga lewisi*. You'll see her, a small winged insect, sitting on top of a gumleaf doing nothing much. But if you look rather more closely you'll notice a long, narrow bubble or blister on the leaf, close to and parallel with the midrib.

If you were to remove the top of the blister very carefully with a scalpel you'd find the sawfly's slender eggs lying side by side in a row. To put them there she cut a series of incisions with her saw between the upper and lower surface of the leaf, making a space for each egg. Tiny serrations along the edge of the blister show where the saw went in. The blister is the leaf's reaction to this invasion. After a while, as the eggs develop inside their secure, temperature-controlled cells, the blistered part of the leaf goes brown. When the larvae hatch they chew their way out.

Still pale and soft, newly moulted sawfly larvae rely on their fellows for protection.

The newly hatched sawflies huddle together in a clump, tails together and heads pointing out. While they remain on the nursery leaf this is their 'between meals' position. The female stands over them, her legs forming a protective cage, buzzing a warning if disturbed. By the time the larvae are independent enough to move away from their nursery leaf, they're on their own.

The big black larvae of the Steel Blue Sawfly, *Perga dorsalis*, are the ones commonly called 'Spitfires'. The adult is a big insect with steely blue-black body and brownish wings. The larvae spend the day resting together as a dark, conspicuous mass in the branches. They're called spitfires because of the oily yellow fluid they 'spit' out when touched. But it's misleading to say they spit, because the fluid just oozes out in a big blob that's swallowed again later. It smells strongly of concentrated eucalyptus, pleasant to the human nose (to mine, anyway) but undoubtedly a deterrent to the parasitic wasps that plague these larvae.

Many people find these insects repulsive, but they're quite harmless, and they have some interesting and quite sophisticated habits. Earlier I called sawflies solitary insects without any social organisation. But as you'll have gathered, the larvae themselves are gregarious. In fact it seems to me that the Gumtree Sawflies take togetherness to extremes.

Right from the start the young larvae huddle together between meals, and even when feeding they don't move far apart. When it's time to moult they cast off their skins supported by their fellows. Look closely at a clump of resting larvae and you might see some pale heads showing up amongst the dark ones. These are the newly moulted larvae that have dropped their cast skins. Still soft and vulnerable, they're safely supported within the cluster.

Then there's the comunication system that the larvae use. When they're not in physical contact they use their tails to tap out audible signals that can bring them together again. There's safety in numbers.

Walking in the bush, you may see a shapeless blob making its way across the ground. Big enough to be a reptile or a small mammal, sluglike in motion and lacking obvious head or tail, it's disconcerting to say the least. Repugnant enough for a shudder of distaste and cries of horror from the uninitiated. But it's just a company of harmless sawflies on their way to a burial.

In order to become adults, sawflies go through a pupal stage like their cousins the ants, bees and wasps. The Gumtree Sawflies do this underground in a communal cocoon. When the time comes, separate batches of larvae may congregate,

ABOVE: *Lewis's sawfly stays for a while with her newly hatched offspring until they are ready to move on.*

moving down the tree as a compact mass. They set out then on a hike across the forest floor that may take days, looking for a good spot. They move as a unit, slowly, in fits and starts. After a stop, one at the 'front end' taps its tail and creeps forward. The rest, in a clump up to 10–15 centimetres across and 2–3 centimetres deep at the centre, tap in return and off they go. Soon they slow down for another breather. Then the tapping starts and they're off again.

Who makes the decision about the final stop? There's no doubt there are decision-makers among them, individuals with more initiative than the rest. When a suitable place is decided on they go underground, still in slow, slow motion. They do it by squirming rather than digging. Once buried they work together to make the 'cocoon', a collection of individual chambers or cells joined lengthwise. Their materials are soil, eucalyptus oil and silk. The end result is a drum-shaped structure with a parchment-like texture, in cross-section rather like a honeycomb.

The adult sawflies take a year to develop and emerge, several years if conditions are not right for them, and they come out one by one at intervals. Life as a huddlesome community is only a dream now as the solitary winged males and females fly off to feed and find mates.

Sawfly larvae adopt typical S-shaped defensive postures.

10 The butterfly that discovered Australia

When stories are told of early wildlife discoveries in Australia it's the marsupials, birds and reptiles that receive all the attention. Nobody mentions the insects. Yet by the time Captain Cook's voyages of exploration ended in 1777, at least 224 different kinds of insect had been collected.

The Wanderer Butterfly (Danaus plexippus) *lays her eggs only on plants of the milkweed family* (Asclepiadaceae).

Among the 37 species of butterfly included in those early collections, the Wanderer Butterfly, *Danaus plexippus*, was missing. And for a very good reason: it wasn't here. The fact is, we never did discover the Wanderer. It discovered us.

This handsome black and orange butterfly that flies with such careless confidence in our gardens and alongside country roads is an American migrant. Back in its homeland, where it's called Monarch, the butterfly is something of a national institution because of its spectacular long-distance flights.

Every autumn the Monarchs take to the wing in great numbers from Canada and northern areas of the United States, and head south. Those that make the journey successfully end up at regular over-wintering grounds in the southern states.

There they gather in huge numbers, crowding together, weighing down the trees and giving autumn colour to groves of evergreens.

Annual festivals are held at some of these locations to celebrate the arrival of the butterflies and the tourists that flock to see them. A much better way to make money out of butterflies, it seems to me, than selling their corpses.

In spring the butterflies head for home again. But very few of them get there. Butterflies don't live long enough to take return tickets for such a journey, differing in this way from migrating birds.

Along the way the butterflies stop to breed near suitable food plants. On these they lay their eggs and then they die. It's the next generation that completes the journey—and who can guess how they do it, without parental guidance?

It was almost a hundred years after the arrival of the First Fleet in Australia that the Monarchs or Wanderers arrived on the Australian continent, under their own wing power, to set up their first colony. By that time white people who preceded the butterflies as colonisers had already adapted and become recognisably Australian. We could say the same about those butterfly immigrants.

In fact some might have felt immediately at home in their new surroundings. Remember the over-wintering grounds in the southern United States, the groves of evergreen trees? Guess what those evergreens mostly are! You won't because it's so unlikely. They're Australian gum trees. Eucalypts are widely planted as ornamentals in the United States, particularly in southern California. What an odd quirk of coincidence that migrant tree and migrant butterfly should first meet on the butterfly's home ground.

It would not be true to say, of course, that the butterflies got a taste for eucalypts and headed for the land of gumleaves. For one thing, Wanderers don't actually feed on gumleaves. For another, the journey to Australia took a long time and several generations. They seem to have island-hopped, the first major stop being Hawaii in about 1850.

OPPOSITE TOP: *Male Wanderers are easily distinguished from females by a dark spot near the inner border of each hindwing.*

OPPOSITE BOTTOM: *Flowers of the introduced* Lantana camara *are a favourite source of nectar.*

ABOVE: *The caterpillars feed on all parts of the milkweed plant, including the inflated seed pods.*

ABOVE RIGHT: *When fully grown, the caterpillar moves away from the milkweed to pupate on a different plant nearby.*

OPPOSITE: *A newly emerged Wanderer Butterfly unfolds its wings.*

Then on to Tonga via the Marquesas and Tahiti, reaching Samoa in 1867. Three years later they were in Queensland, all set to colonise the whole of Australia's east coast and west as far as Adelaide.

The domestic migrations of Wanderer Butterflies must have been going on for thousands of years. Many must have crossed the American seacoast and even made island landfalls, but failed to survive. So why, all of a sudden, over a brief 200 years, have they succeeded in colonising most of the temperate parts of the world?

You could say the butterflies were helped by a bug, the travel bug. It was around the middle of the last century that European man started his migration explosion, setting up colonies in faraway places, establishing new trade and travel routes. And wherever he set foot on land he spread the plants of the Old World and the New, either accidentally or on purpose. The special food plants of the Wanderer or Monarch Butterfly were among them.

These butterflies feed on plants of the Asclepiadace family, the ones known variously as milkweed, cottonweed, kapok plant or swan plant. The name milkweed comes from the milky fluid in leaves and stems, which you will know about if you've ever stopped by the roadside to pick the tall sprays of white or red and yellow flowers. In contact with the skin this fluid can cause severe itching in some people.

The milkweed fluid is toxic to birds and mammals but does no harm to the caterpillars of the Wanderer that feed on every part of the plant. In fact the poison is carried over from the caterpillar stage through the pupa to the butterfly. So the butterfly, a nectar-feeder, retains the immunity of the caterpillar. No wonder the Wanderers float about our gardens with such an air of insouciance, thumbing their noses, as it were, at avian butterfly-collectors.

Milkweeds are native to America and Africa but few other places. Oddly enough, the most common milkweed our Australian Wanderers feed on comes from Africa.

So two migrants, a dainty plant and a handsome insect, have come together from opposite sides of the world to grace our Australian gardens and roadsides.

11 Emus think we're funny too

Male Emus grunt and rumble. It's the female that makes those distinctive sounds like water glugging out of a bottle, usually referred to as 'booming'. I don't know what this says to male Emus but I find it as evocatively Australian as the notes of a didgerydoo.

And there's only one other sight as classically Australian as a leaping kangaroo. It's party of Emus, second-largest birds in the world today, stalking in single file across the blue-green plains of the outback.

The Emu's scientific name is *Dromaius novaehollandiae*, which means 'swift runner of New Holland'. Although Emus going about their business in the wild move in slow and stately procession, they can go almost as fast as a kangaroo, clearing three metres of ground at a stride. My colleague Jim Frazier once clocked an Emu pacing his car at around 70 kilometres per hour for a short stretch.

Wild Emus often panic at the approach of a vehicle, running ahead of it even when there's an obvious escape route. When they run, their heads stay still in relation to their bobbing bodies, except to swivel from side to side in order to keep an eye on their 'pursuer'. Usually one bird has the sense to turn aside and then the rest follow.

But wild Emus can also be inquisitive. If you stop your vehicle you can be lucky enough to have the fleeing flock stop, too, and come over to check you out. This happened not long ago when Jim Frazier and I were filming on location at Kinchega National Park in far western New South Wales.

We had stopped the car to scan the trees on our left for nesting birds. As we approached, a party of feeding Emus took fright and disappeared over a sand ridge on our right, and we didn't expect to see them again. But suddenly a long neck topped by a flat Emu head rose up against the skyline, then another, and then the whole flock of emus all came down the sandy slope together to where we sat in the car.

Everyone has visited a camp-site or picnic ground outback where tame Emus come poking around, grabbing the barbecue chops, greedy for anything that's offered, delighting everyone with their funny faces and amusing ways. But these were different. These were no scrawny scroungers, comic characters about to sell their dignity for a handout . . . Emus are not allowed to be fed at Kinchega. These were well-muscled, healthy birds standing tall and proud, shiny plumage lifting and falling as they stepped, setting down their three-toed feet as precisely as any ballet dancer.

Even as the birds milled about, advancing and retreating, curious about us yet still hesitant to come too close, there was a confident freedom about them that you don't see in tame or captive Emus. I found them quite unexpectedly beautiful and watched with regret their eventual bouncing retreat over the rise and out of sight.

A family party of Emus may comprise young ones with both parents or with only the male parent. As with other flightless birds like the Cassowary and the South American Rhea, all parental duties fall to the male bird. The female drops her responsibilities with her eggs, usually eight or nine. Sometimes two females lay their eggs together. This may be necessary because they share a single mate-cum-incubator.

Breeding takes place in the cool months. After mating, the female returns to her life of leisure. The brooding male covers his charges like a warm, shaggy rug, seldom leaving them for eight weeks,

JIM FRAZIER

Australia's largest bird, the Emu (Dromaius novaehollandiae) *shares the outback plains with the Red Kangaroo, our largest furred mammal.*

ABOVE: *The male Emu broods the eggs, covering them like a warm, shaggy rug.*

hardly feeding. He lies low, feathers blending with the leaves and grasss, head and neck stretched out along the ground. You could easily trip over him. I've seen a sitting bird pick up sticks and leaves and drop them back over his shoulder, adding to the natural camouflage.

The dark green eggs are so big the Aboriginal people once used the shells as water containers. They must make an uncomfortable bed for the male Emu, but he is most delicate in his handling, or beakling, of them. If he fidgets, and an egg rolls out, he uses his beak to scrape it back under cover.

Emu eggshells are tough. A friend who has had many breeding birds in his charge believes the male bird himself makes the initial crack. Alerted perhaps by the sound of cheeping from inside an egg, he strikes down on the egg with his breast-bone until the egg cracks in several places. After that, it's every chick for him- or herself.

The striped emulets are handsome. They flop out bleary and bedraggled, but in no time they' re running around their father, brown eyes alert and shiny. The male preens and beakles them constantly. When the eggs are all hatched he leads them away to join the flock, or their mother(s), but at night they still hide from danger and cold under the paternal eiderdown.

ABOVE: *The first egg starts to hatch.*

The first sighting of Emu tracks by a white man was recorded by the Dutch explorer Willem de Vlamingh in 1697 on the Western Australian coast. The first detailed description of the bird itself came from the pen of John White, over on the opposite side of the continent, and that was nearly a hundred years later.

White noted in his diary of February 1788: 'A New Holland Cassowary was brought into camp.

The male bird will care for the chicks until they all move away to rejoin the flock.

This bird stands seven feet high . . . and in every respect is much larger than the common Cassowary . . . '. He reports the bird to be 'not uncommon in New Holland, as it has been frequently seen by our settlers both at Botany Bay and Port Jackson. . .'.

For a long time Emus were confused with their cousins the Cassowaries, already known from islands to the north. The later name Emu was originally spelt Emeu. It is not an Aboriginal name but came from the Portuguese Ema, meaning crane or generally any large bird.

Aborigines had always hunted Emus for food. But it was the white man who became the Emus' most ruthless predator. From the start early settlers killed them for food and oil. Later they were hunted for sport, and because they competed for human food crops planted in their feeding grounds.

It was earlier this century that we hunted the Emu most ferociously. In the 20s and 30s they took the blame for an embarrassing and costly blunder on our part. The birds were blamed for the spread of introduced Prickly Pear (*Opuntia* species) that invaded millions of acres of rural land, making it useless. Over two years the Prickly Pear Commission of Queensland organised the slaughter of more than 100,000 Emus, and the destruction of roughly the same number of eggs.

I wonder how the birds were supposed to have spread the pest. Did they eat the seeds? Did they carry bits of the prickly stuff around between their toes? The records don't say. But here's a comment from birdman Alec Chisholm's book *Bird Wonders of Australia*: '. . . while that slaughter was proceeding, an entomologist who examined the stomach contents of a single emu found there 2991 injurious caterpillars.' Similar records have been made in other areas. What the authorities were doing was killing off one of their best allies against insect pests.

In the same book Chisholm notes that Emus will eat practically anything, from caterpillars to old boots. While Emus in the wild keep their distance and their dignity, the tame ones in zoos and sanctuaries tend to turn silly and want to nuzzle us. It's just cupboard love of course—they've come to think of us humans as mobile, beak-in-the-slot snack-vending machines.

On the whole Emus are gentle and peaceable birds, but they carry an effective weapon in those large, almost reptilian feet that can move daintily yet kick the dust up at 70 kilometres per hour. If you ever do see one racing towards you with a less-than-friendly look in its eye, don't panic. Stand your ground with one arm raised high above you, wrist bent, hand pointing like a beak. This will turn you into a Super Emu and your assailant will dodge past, pretending it was just out for a jog.

Alternatively you could offer it a caterpillar or an old boot.

OPPOSITE TOP: *The stiped emulets are handsome and alert.*

OPPOSITE BOTTOM: *Emus are as much at home in tropical savannah woodlands as they are on the arid plains of the outback.*

12 The huntsman is a lady

Those big hairy spiders that come indoors on summer nights and sidle across your wall or ceiling strike terror into many hearts. When the occasion arises, when you need to call your spouse or significant other to deal with the situation, how do you refer to these unwelcome invaders?

Large Huntsman spiders (Isopoda vasta) *are often found indoors, but they're harmless to humans.*

No, seriously, apart from ugly, dirty, dangerous horrors, which they are not. Do you call them Triantelopes, Tarantulas, Huntsman Spiders, Sparassids or Funnelwebs? The ones I mean are

LEFT: *Huntsman spiders come plain and fancy; this is a female* Isopoda insignis *with her eggsac.*

the rather flat-looking spiders with long hairy legs and back-to-front knees. When you poke them they tend to come abseiling down from the ceiling or walls with great speed and go galumphing across your carpet. Perhaps galumphing isn't quite the right word, because my dictionary says that means 'to march exultantly with irregular bounding movements'. The spiders don't do it exultingly. Those erratic movements are simply evasive tactics. But getting back to names. The last one first. If you call them Funnelwebs you're obviously an arachnophobe. Funnelwebs don't climb walls, they just send you up them. If you call the spiders Sparassids then you're probably the opposite, an arachnophile, that is, a 'spider freak' who knows all about them. Sparassidae is the scientific name of the spider family to which these belong. There are lots of spider families. Funnelwebs belong to a different one.

The Greek word *sparasso* actually means 'to rend apart'. But is this really what spiders do to their prey? We say, 'The Greeks have a word for it', and then often use quite the wrong Greek (or Latin) word when we, or rather biologists, make up a scientific name.

The truth is that in spite of their fearsome-looking fangs, spiders don't rend their prey apart at all. They pour digestive fluid over their victim, mash it about a bit then suck it dry when the inside has turned to pulp.

If you call the spider in your house a Huntsman, well, that's the name used in most English-speaking countries where they live. A good enough name, although most of the Huntsmen you see are females. Politically correct Huntsperson seems a little clumsy. Those other two names should be dropped. Tarantula is confusing, Triantelope simply meaningless. In his book *The Australian Language* Sydney J. Baker tells us that people used Triantelope at least as early as 1835. The word seems to be a purely Australian corruption of Tarantula. That's where it gets confusing, because Tarantula is a name the harmless Huntsman doesn't deserve.

But the confusion started long before Australia was even heard of. The original Tarantula bore no relation to our Huntsman, or anyone else's. In the Middle Ages the people of an Italian town called Taranto came down with a strange new sickness. It was said to be caused by the bite of a species of

Hanging helpless from its old skin, the spider is most at risk just after moulting (Isopoda vasta).

Like all spiders, a Huntsman digests its insect victims externally.

Wolf Spider that had suddenly appeared in great numbers in the fields where people worked. The malady later came to be called *tarantism* after the town, and the spider to be called Tarantula after the town and the malady.

The supposed cure for those affected was to get together in public and dance in a frenzied and abandoned fashion until they fell down exhausted. Or perhaps just lay down. Because there's another explanation for this strange occurrence which, if true, exonerates the spider completely.

It seems the outbreak happened at a time when church authorities had banned dancing and cavorting in public and all such sensual goings on. Well, the story goes that the people of Taranto went on with their high jinks without benefit of clergy, as you might say, and used the harmless arachnid as a *scapespider* . . .

So much for tarantism. It certainly gave spiders generally a bad press. In newly discovered America some big, hairy ones were also named Tarantula, after the supposedly horrific Italian. In fact the two are unrelated. The American one is a cousin of our Funnelweb. Our Huntsman is related to neither.

Huntsman spiders come plain and fancy, though nearly always in shades of brown or grey. But look at the green *Olios* spider in the picture, one of the prettiest around. It's a small kind of Huntsman, a member of the same family. Other kinds of Olios are mostly brown, and they're

ABOVE: *A smaller member of the Huntsman family is this species of* Olios.

ABOVE: *Huntsman spiders, like this* Isopoda vasta, *lie in wait for victims on frequently used insect trails.*

smaller and fatter than true Huntsmen. They do their hunting around foliage and leaf litter.

To encourage goodwill towards spiders I usually tell people they come indoors to hunt flies and other pests. Well, so they may. But it's more likely they simply use our houses for shelter. To a Huntsman a house is probably just a huge hollow tree or cave occupied by strange two-legged animals.

If you want to do a Huntsman a favour when you find it in your house, just wait for it to sit still then pop a jam-jar over it. Slide a piece of cardboard or stiff paper gently underneath, carry it carefully outside and set your captive free by shaking the jar over a bush.

If you go out hunting for a Huntsman with a torch on a warm night the most likely place to find one is on a gumtree. Tree trunks are busy highways for insects going aloft to find nectar or other insects to prey on. The Huntsman is best described as a highwayman, lying in wait with outspread legs, ambushing whatever takes its fancy. Any insect on its way to the top must face many such hazards. But the aggressor doesn't always come off unscathed. Notice the drop of greenish fluid on the feeding spider in the photograph. The tree cricket's spiky legs or powerful jaws drew blood before she succumbed.

If you're wondering why that spider isn't on a tree trunk after what I've just said, well, spiders are individuals. Some Huntsmen like a change. You can usually find one or two lurking behind the strap-like leaves of such garden plants as Clivea.

A huntsman creeps easily into crevices and small spaces, partly because it has such a flat body and partly because it bends its 'knees' horizontally instead of up and down. It is designed for hiding behind the bark of trees during the day.

Sometimes if you pull a piece of loose bark away from a tree you'll find a female Huntsman hugging her white egg-sac, guarding it against egg thieves. Newly hatched spiderlings, speckled and fragile, stay with their mother for a while, feeding around the edges of any morsel of food she is holding or regurgitating. Then one by one they drift away on spindly legs to find their own place in the world.

13 Seasons in the sun

It's joy and fun, and light at the end of a long, dark tunnel for cicadas, come summer time. No wonder they sing fit to burst our eardrums. Wouldn't you, after boring years underground wearing damp clothes in the damp earth? No company but worms and woodlice, nothing to do but suck on roots and dream of flying?

ABOVE: *A newly emerged Red-eye cicada* (Psaltoda moerens) *shows fleeting beauty.*

In the eastern states, from coast to mountains and beyond, it's been a very good year for cicadas. The best for many years, they say, in terms of decibels. Male choirs led by Cherry-Nose and Double Drummer, Red-eye and Green Monday, Floury Miller, Squeaker, Washerwoman and the rest, came together for an insect eisteddfod on a grand scale. The trees trembled, the air throbbed. It was a noisy celebration of light, life and love in the sunshine.

Some among their captive audience might say it was a bloody awful year for cicadas. A revolting din. An invasion of privacy that ought to be legislated against. Well, we grumble a bit, but cicadas are a part of the blue and green and golden summers that we wouldn't want to be without. For people of the east coast towns and cities they're as Australian as a ute-load of mallee roots. We wouldn't have given them those fancy names otherwise.

It was a good year not only for cicadas, but for insects of all kinds, with the drought undeniably at an end. Outback, colour flowed like brushstrokes on a bare brown canvas. Everywhere, trees and shrubs re-opened or expanded their factories. In flashes of colour and whiffs of scent the word went around: food's on!

It didn't take long for the insects to get the message and soon the land was a-buzz, a-leap and a-chomp with them. On the ground, ants gathered bumper seed crops. In the air, butterflies breezed from flower to flower, picking and choosing. Among the leaves, caterpillars conscientiously recycled nutrients, harvesting with one end, spreading fertiliser with the other. And up in the gumtrees, cicadas sucked and sang.

For city-dweller, which most of us are, such insect population explosions are more often heard than seen. Bugs on bushes may be overlooked or ignored, but you can't just switch off your ears.

OPPOSITE: *Blue wings like rippled glass will lose their colour by morning. Green Monday cicada* (Cyclochila australasiae).

Cicadas are pretty hard to see in any case. I've had more than one cricked neck trying to get a fix on calling cicadas. Those bulging eyes are not just part of a pretty face, they can see you coming. And cicadas have a sneaky way of sidling around the back of a tree trunk when you get too close. This makes them sound more distant, so you walk past to check out the next tree. Then they sidle back again.

If you stand too close to a tree full of cicadas they can give you a shower bath. Not intentionally, it's just part of their feeding behaviour. All the time cicadas are calling they're taking in sap from the tree, using a built-in sucking tube with a sharp point at the tip and a small but powerful pump at the base.

As it takes a great amount of sap to provide only a few calories for the insects that have no other food source, there's a lot of sucking to be done. This means there's also a lot of waste fluid to be squirted out. Nothing unpleasant or smelly, but it's a little disconcerting when there isn't a cloud in the sky and suddenly it's raining.

While they're underground the young, wingless cicada nymphs drink sap from the roots of trees in the same way their elders above drink it from the branches. The fluid they squirt out turns the soil into mud. As the nymphs grow they need to enlarge their burrows. They do it by plastering the mud against the tunnel wall, filling the air spaces and consolidating the walls. That's why you never see a pile of soil heaped up around those neat cicada holes in your lawn. And that's why those

OPPOSITE: *Stages in the transformation of a cicada after leaving its underground burrow* (C. australasiae).

ungainly little monsters that emerge from the ground for their final moult often look as though they've had a mud bath.

It's the empty shells of those muddy troglodytes we mostly see in the cicada season, hooked up by the claws on trees and fences. They are evidence of the final ecdysis, the hours-long striptease that signifies a cicada's coming-of-age. This year I counted 70-odd shells on and around the base of a single roadside tree, and every nearby bush and tree had its quota. This was Double Drummer territory. Earlier emergents were in full swing above, but others were still hanging perilously head downward, extracting legs and wings.

Double Drummers come out of their shells in full daylight, so it's easy to watch them. Easy for predators to pick them off, too, and perhaps that's why so many come out together. A bird can eat only a few at a time. Most of the other big cicadas, such as the Greengrocer and Yellow Monday and Cherry-Nose, do it at dead of night.

The creature that emerges from the splitting shell is unlike either its former or its future self. Out of all the years below ground and all the weeks above, for one night only, and for just a few hours of that one night, an insect Cinderella becomes a princess.

Not without a lot of trouble, though. It might seem easy to the casual watcher. But there's more discarded than the rigid outer cuticle, and I for one prefer to shed my skin the human way, little by little, in unnoticed, powdery fragments. Here's why. If you look closely at a moulting cicada you'll see white threads and strings linking the old skin and the new. These are nothing less than the linings of the body openings, already replaced by new ones.

Imagine it! All that unwanted inner tubing from the hind end of the gut, from the little breathing tubes that open like portholes along an insect's body, from the mouth and gullet; all this must be withdrawn, very slowly, perhaps painfully, as the insect leans backwards out of its shell. Think of *that* next time you have a sore throat.

But it's the colours of this reconstituted insect that catch the eye. The contents of that unlikely brown package come in delicate opalescent tints, plated with gold leaf or beaten copper, or tinted with touches of tangerine. Then there are the wings, unfurling from the shoulders like draperies at first, flattening into panes of blue or lilac ripple glass, finally folding tentwise ready for use.

By the time the cicada takes off for its first flight in the morning sunshine the new skin has hardened and darkened, and the wings become clear. All a Red-eye has left of its fancy dress is the ruby-red eyes; the rest is black. Some, like the Yellow Monday, are still pretty enough. But most of that delicate beauty seems to be just a side-effect, a mere

ABOVE: *The Red-eye cicada, now drably clad, sucks sap from the stem of a gumtree.*

TOP: *A Yellow Monday cicada* (C. australasiae) *lays her eggs in slits along the branch.*

A Double Drummer cicada (Thopha saccata) *spreads and rustles its wings in a defensive display.*

accident of insect chemistry that's of no significance in itself, only to human eyes and sensibilities.

Because of their singing, cicadas have been familiar all down the ages in many countries of the world, but I doubt that any other people have given their local species such delightful names as ours. Perhaps the Chinese did. They used to keep cicadas indoors in cages, like songbirds. I wouldn't like to live so intimately with one of our male Double Drummers in the mating season.

The Americans have their 17-year cicada, of course. They call it simply the Periodical Cicada. The scientific name, *Magicicada*, is much better. The amazing thing about these Periodical Cicadas is not just that they live 17 years underground but that they all come up together in the same year. If some of the nymphs get on to a better food supply and mature before their fellows, well, they just have to kick their heels down there until they hear the starter's signal, whatever that may be.

We don't know much about our own Australian cicadas' life span underground. Some are thought to take five years or more to mature. Whatever the time span, sooner or later some mysterious directive sends them all clambering upwards at the right time, out of the soil and onto the nearest vertical surface. The life they prepare for there lasts no more than a few weeks.

The final moult of a cicada nymph is an act of magic, deserving a flourish of trumpets, spotlights and an audience. Yet it happens quietly and unobtrusively, seen by few save the odd wandering naturalist like me who photographs the whole affair so you can share it at ease in your armchairs. More comfortable than hunching down in the dark on your keees for hours at a time.

But it's not quite the same experience as seeing the live performance, and here's one insect voyeur who goes along season after season to watch the repeats.

LEFT: *Its cast-off skin secured to a tree trunk, a Double Drummer cicada comes of age.*

14 It's a frog's life, down-under...

Everybody knows about marsupials: kangaroos and koalas and all those other quaint down-under animals, most of them charming if not downright cuddly. But who's ever heard of a Marsupial Frog? I'm not joking. Marsupial frogs really exist and like their furry namesakes they carry their offspring around in a marsupium, or pouch. But in this case it's the daddy that does the carrying.

The Marsupial Frog is tiny and quite nondescript to look at, and it lives on the rainforest floor along the mountainous Queensland–New South Wales border. Biologists are still studying details of its astonishing breeding behaviour, but the broad facts are now well known.

Instead of a single pouch like a kangaroo the frogs have two of them, one on each side like a pair of hip pockets, and it's only the male that has them. He goes round for weeks in summer looking pregnant, with his pockets stuffed full of tadpoles.

So how do the new hatchlings get into the pouches in the first place? We know how newborn kangaroos do it, hauling themselves hand over hand up their mother's belly, gripping her fur with specially developed front limbs. But a newly hatched tadpole doesn't have much going for it except a tail. And these hatch out as little more than embryos; just a lot of blind white squiggles in a mess of jelly.

Well, the male squats down right in the middle of the egg mass when the eggs are about to hatch. And somehow, squirming about all over their father's slippery body, the tadpoles emerging from the eggs find the narrow slits and force their way in head first. There they stay safe and soggy to pop out again in a few weeks later as fully formed froglets.

Only a few years ago an even more astonishing example of parental care came to light. It's called gastric-brooding, which means *rearing your offspring in your stomach!* In this case the caring parent is the female. She gives her young ones the best possible start in life—by swallowing them.

But this is paradoxical, you'll be thinking. Surely they'd be killed by the acid gastric juices, torn apart by the turbulence, swept downstream by peristaltic action—in short, digested ?

They would indeed if it were not for one astonishing fact. When the embryos of the Gastric-brooding Frog enter their mother's stomach they throw a chemical spanner into the machinery of digestion. Everything comes to a temporary halt. No more acid production. No more churning and burning. For the female, of course, no more feeding for a while.

And so the tadpoles, fed from reservoirs of yolk stored in their own bodies, grow all the way to froghood in the safe and tranquil pool of their mother's *milieu interieur*.

This is a breeding strategy that's unique not only among frogs but in the whole animal kingdom; not only in Australia but throughout the world. But the Gastric-brooding Frog, found only in a restricted area of south-east Queensland, is very rare indeed. This is a sad thing because its juvenile chemistry could hold the key to a vital breakthrough in the treatment of stomach ulcers. Imagine being able to close the factory down completely while repairs are carried out!

Most frogs don't seem to care a croak about the future of their offspring; they just lay them and leave them. Or so it seems to us when we see a mass of tadpoles struggling for a living in a drying pool.

But in fact even the most careless-seeming mother frog has an instinctive awareness of her offsprings' future needs. Herself an air breather,

A froglet emerges from its mother's stomach several weeks after being swallowed alive (Rheobatrachus silus).

DALE CAVILLE

ABOVE: *A female Corroboree Frog* (Pseudophryne corroboree), *with her large eggs in a Snowy Mountains sphagnum bog.*

TOP: *Corroboree frog* (P. corroboree), *named after painted designs used in Aboriginal rituals.*

she lays her eggs in or near water, essential for her gilled tadpoles. Furthermore, she may provide protection from the elements and from enemies by surrounding her eggs in jelly or embedding them in a floating raft of froth.

Today, though tropical and temperate rainforest still remain, it's the great tracts of temperate eucalypt forests that are home to the dominant group, the tree frogs or hylids.

But, although pocketing or swallowing your pride and joy may be exceptional, it's really quite common for one or both frog parents to take a continued and even active interest in their families.

Take the Corroboree Frog—and as frogs go you couldn't take a more beautiful one. *Pseudophryne corroboree* was named for the likeness of its striped pattern to some of the painted designs used in Aboriginal rituals. But those were done in earth pigments of sombre colours, while here it's a high-gloss enamel job in black and vivid orange-yellow. But these striking little frogs are secretive, and few get the chance to admire them

Corroboree Frogs are found (not easily) above the snowline in Australia's Southern Alps. There's no snow there in summer and the high sphagnum bogs where they live resound to the curious, creaking territorial calls of the males.

The frogs tunnel in the moist, spongy layer that lies under the green carpet of sphagnum moss, and down there in a nest chamber above water level the female lays her large eggs. Both parents stay with the eggs, guarding and attending them.

If it rains and the water rises high enough, the eggs are submerged. If necessary the parents give them a nudge or two in the right direction. Then the tadpoles develop in the usual way. But if it doesn't rain, nothing's lost. Embedded in the jelly-like fluid of the eggs, the tadpoles are still able to develop nearly all the way to froghood. They even get some of their adult colouring in there.

It's surprising to find that the inventors of portable-parental-puddles and development-within-the-egg actually live in the favourable conditions of wet forest, permanent creeks and snow-fed bogs. Yet the outback frogs, with their enormous water problems, have no such sophisticated forms of child care. Out there, though, it's not so much the tadpoles as the frogs themselves that need special survival strategies.

Australia is the Dry Continent: four-fifths of it arid and all of it subject to drought. Most frogs

RIGHT: *Tadpoles gasp in sunbaked puddles, fighting for food and space.*

(the hylids) live in the higher rainfall areas around the coast. Yet, amazingly, there's hardly any part of the dry hinterland where you won't find frogs. So how do they manage? A frog must be moist at all times. A tadpole out of water is a dead tadpole.

In the outback there's little or no permanent water and rain is unpredictable. Years of drought alternate with widespread flooding. When there's water there's plenty of it, but it soon dries out. Frogs must be opportunistic breeders, and development needs to be swift. So the eggs laid in times of flood hatch quickly, tadpoles grow at high speed, and it all becomes a race against time.

As the floodwaters evaporate, the tadpoles gasp in sun-baked puddles, fight for food and space. Many starve, some may turn cannibal. Predators take a huge toll. But enough survive, and as frogs they have the drought to cope with because as sure as night follows day, drought follows flood.

Perhaps uniquely among wildlife, outback frogs benefit from man's presence among them. The frogs move in to share dams and troughs and bores with sheep and cattle. They take up residence around the leaky taps of homestead water tanks, among pot plants on shady verandahs. Where there's internal plumbing they lurk behind WC cisterns—and what more permanent pond than the one in our bathrooms, even if it's not entirely ideal for breeding.

But if you think humans are the only animals that can catch the rain and store it against dry times, think again! Some of the frogs out there, the real outback diehards, do it as a matter of course. Using only their bodies as containers, they can lay up enough water in one wet season to see them through years of drought.

Although they need plenty of it and easily dehydrate, you'll never see a frog lapping up water.

One of several large tropical tree-frogs common around houses and gardens in north Queensland (Litoria infrafrenata).

In the arid outback, frogs take up residence around leaky homestead taps (Litoria aurea).

A desert frog doesn't get many chances to sit in a puddle and reflect (Neobatrachus centralis).

Instead of drinking they absorb moisture through their skin. The ones that live near water have very smooth skin. The granular skin of a dry-country frog gives it a much greater area for absorption, essential where evaporation is rapid and rain infrequent and irregular.

The desert water-holding frogs go further. They have a set of sac-like spaces under the skin that can be filled up like water bags. They have a water-storing bladder that will stretch to an enormous size without discomfort.

At the end of the wet season, when the transient outback flowers wither and the soil starts to crack, these frogs, well-fed and with their breeding done, go into their personal survival routine. First, a good long drink.

Lying in a puddle or a mud hole or just flat out on a patch of damp ground each frog absorbs as much water as it can, mostly through the skin of its underside. The sacs and the bladder inflate, the water penetrates between the very cells of the body until the frog is virtually waterlogged. It swells up, round as a ball, with tight, shiny skin.

So far so good. The water supply is now secured. But what goes in so easily can come out the same way. Water evaporates through the skin and in the drying air dehydration is rapid. So what does the frog do next?

It doesn't do much at all, it just sits in the mud gazing into the distance. Behind it, though, there's a slight movement. The mud's stirred up a bit and the frog seems to be sinking . . . sinking . . .

Before disappearing underground, a burrowing frog raises transparent lids to protect its eyes (N. centralis).

Before it disappears from sight, pick it up and have a good look at the soles of its feet. Not its hands. (Isn't it odd that we use the terms *hands* only for frogs and primates while other animals have 'front feet' or 'paws'?) On the inner side of each foot you'll see a large, horny, scoop-shaped tubercle. These are the shovels that can bury a frog many feet underground.

Put the frog back in the mud now, and watch the rear end. Left, right, left, right—the hind feet shuffle the mud out sideways, making a depression into which the frog slowly sinks, still gazing into space with its high bulging eyes, until the mud closes over them.

Just before the frog goes under you'll see something odd happen to those eyes. Over each one, not necessarily at the same time, a solid white stripe moves up from the bottom. It's the upper edge of the transparent lower eyelid, a protective shutter against the abrasive soil.

Much has been learned about burrowing frogs from specimens dug out of the ground and set up in artificial observation cages. Over the years, while filming them in special glass-fronted containers, cameraman Jim Frazier and I have been able to follow stages of their sojourn underground. This is what happens.

We place the frog, well-fed and well-watered, into the narrow container filled with damp sand or mud. We cover the glass front with a cloth to simulate underground darkness. The frog doesn't take long to burrow downward. Once there it makes a roundish chamber for itself, pressing and consolidating the mud walls by moving about. Then, folding its hands and feet close to its body, it goes into a dormant state. So do we, for a few weeks, at least as far as frogs are concerned. Meanwhile, a remarkable thing happens.

If you haven't seen this before it comes as a surprise to lift the cover and find the frog done up like a frozen chicken in a plastic bag. What happens is that while it lies motionless a transparent, impervious skin forms around the frog, enclosing it tightly, eyes and all, in a sort of cocoon. Only the nostrils remain connected to the air, via two tiny tubes. This is the way the frog conserves its water supply.

Out in the desert, as time goes by and the sun beats down, such a frog remains cool and moist in its cocoon, appearing asleep, waiting out the drought. Around it the soil may dry rock hard. Then one day, perhaps only months later, perhaps years, the drought breaks once again and rainwater seeps down into the soil.

ABOVE: *'Plastic wrapped' like a frozen dinner, a waterholding frog may survive drought for ten years or more* (Cyclorana cultripes).

So we simulate the end of the drought, gently pouring water into our container until the soil is saturated. When the water reaches the frog the cocoon softens, the frog wakens, stirs, opens it eyes under its wrapping.

Now with its hind legs the frog reaches up and back, clawing at the flexible overskin no longer needed, wrinkling it forward as if to pull off a glove, but instead stuffing it all into its mouth. Gasping and gulping, using feet and hands and not without difficulty, the frog gets it all off and swallows the lot making horrible faces in the process.

Now it burrows upwards and joins its fellows popping up everywhere out of the ground into the glorious rain, ready for a short, hectic season of feeding and breeding.

For the burrowing frogs of the desert, life is a manic thing, an uncertain pendulum swinging between hyperactivity and inertia. But success is measured by survival. It's been a long, long time since the lush rainforests and broad waterways of Australia's ancient hinterland gave place to arid plains and deserts and the frogs have made the necessary adjustments.

Although their ancestors were the first back-boned animals to step out on to dry land, frogs still seem to live, as it were, with one foot back in the ancestral waters. The word amphibian says it all: *amphi* and *bios*, a 'double life'. The adult frog air-breathing, carnivorous and mobile on land. The tadpole gill-breathing, vegetarian and still totally dependent on one kind of puddle or another.

But who am I to talk? When you think about it, aren't we humans ourselves aquatic for the first nine months of our lives? In that safe and tranquil pool . . .?

When rainwater reaches its underground cell, the waterholding frog tears off the softened 'cocoon' and eats it (C. cultripes).

15 Mountaineering moths follow ancestral trails

GLEN CARRUTHERS

Bogong moths (Agrotis infusa) *migrate annually from the dry plains to spend summer in the mountains.*

I'm standing on a granite outcrop high on the westward side of a remote ridge in the Southern Alps, waiting for the start of an 'insect spectacular'.

Ahead, the snowless blue ranges stretch like an endless petrified sea, backlit by the setting sun. Somewhere below, hidden among the snow-gums, a late currawong calls sweet and strident. Wind whistles harmonics through crevices in the rocks around and below me. Any minute now I'm expecting something more than wind to come out of those crevices.

From a vantage point down among the boulders my colleague, Jim Frazier, has his camera set up and he's waiting for my signal. We're about to film the climax of a film sequence, part of a natural history documentary we're making for the BBC's TV series, *Natural World*.

The stars of the film are all Australian moths, each with its own fascinating story. The film, which I've called *Desire of the Moth*, deals with the complex drives that motivate these moths throughout their highly organised lives.

Many of these drives are explicable in human terms, like the urge to find a mate and reproduce, but others are more enigmatic. The attraction of moths to light, for instance, while well known, is not well understood; old theories favour navigation as a basis, new ones favour sex. Whatever inspires it, the poet Shelley called this often suicidal preoccupation with light 'the desire of the moth for the star', which gave me the title of our film. But the urge we're concerned with up here among the granite boulders of

ABOVE: *The moths have traditional camps among granite outcrops of the Great Dividing Range.*

Australia's highest mountains is a different one.

On the map of this vast area the name 'Bogong' occurs over and over: Bogong Peak, Mt. Bogong, Bogong Ranges, Bogong High Plains. All these places were given their names by Aboriginal people, and from the number and obvious importance of the sites you'd never guess they were named after a very small and quite ordinary brown moth. To the Aboriginal people the moths, called Bogongs, were once a rich annual source of food. To get them they had to climb mountains and ransack the caves, cracks and crevices in the granite outcrops where the moths congregated.

In its caterpillar stage *Agrotis infusa* is a lowlander, a crop-chewing pest of pastoral and farm lands, one of the 'cutworms'. As the pastures dry out in spring the winged moths emerge from the ground where the caterpillars have pupated. Their bodies are fat with stored fuel. They'll need it. They're about to set off on an incredible journey, following ancestral flight paths to the mountains of the Great Dividing Range. Though heavy bodied the moths are not large and their wingspans are limited. Yet by the time they arrive some of the moths will have travelled 1000 kilometres or more.

From thousands of square kilometres to the north and west the travellers set off, converging in ever-increasing numbers. Along the way they stop at traditional staging camps, resting for a while then going on in company. Since white men came and built settlements across their flight paths many moths have ended their journey prematurely, around street lamps and windows, swept and vacuumed out of houses.

In a small mountain town in New South Wales a chipboard factory lies directly on the site of such a staging camp. The moths use timber stacked in the yards as a shelter, swarming in dusty clouds when disturbed. They roost inside the mill, in

GLEN CARRUTHERS

Jim Frazier prepares to film the evening emergence of the moths from rock caves and crevices.

Where the moths gather, predators like this carabid beetle flock to feed on them.

every dark spot, in every bit of machinery. They even roost in the rollers that squeeze out the sheets of chipboard. Discards pile up daily as the sheets are scanned for pressed moths that look like instant fossils, not yet exploited as the latest in domestic decor.

People, towns and chipboard factories notwithstanding, millions of moths make it annually to their destinations: long-established sites on certain westward-facing granite-topped ridges and summits. The moths creep on arrival into the cool, dark spaces between boulders, lining the walls head to tail like furry brown tiles. By midsummer such a tumble of rocks will house thousands upon thousands of them. Below them on the ground a mushy mess of droppings, and the crumbs of the feasts of predatory beetles, prove centuries-long occupation, and that is probably a conservative guess.

The moths are aestivating—the summer equivalent of hibernating—but they're not dormant. They don't breed here either; they do that when they go home again in the autumn. What they are mainly concerned with up here seems to be their nightly ritual dance against a backdrop of stars.

In times past the Aboriginal people came up these mountains for an annual feast on the Bogong Moths. They crept into the crevices, filling woven bags with their booty. They threw the moths onto hot ashes to remove wings and scales. It was a time of plenty. Now only a few white people know of the sites, and only a handful of these have seen what we are about to film.

During the day the Bogong moths stay hidden, lining the rock faces like tiles, head to tail.

GLEN CARRUTHERS

The disc of the sun has slipped under the horizon. The rocks are in shadow. Now watch! A few dark bodies scamper out of an exit hole and take off. More follow, running over the rocks. Soon the moths form a dark stain around cracks and crevices as they boil up between the boulders.

Jim's camera is whirring, using the last light of the afterglow to film the exodus. He moves between two granite tors to get a shot of the flying moths, just black silhouettes in whirling motion against an orange sky . . .

Now there's only starlight and a rind of moon. The main mass of moths is out and on the move. In unbelievable numbers they sweep and gyrate above the rocks, collide in mid-air, the choking 'dust' from their wings filling our mouths and nostrils. And still they come, pouring like smoke in direct take-off now, straight out of the spaces between the boulders.

Jim has the portable movie lights rigged up, and the flying moths now shine white on black. He films them fast to slow them down in the finished film. The eye as well as the camera plays tricks. When I try to follow the flight of individual moths I get an after-image that gives me five moths for every one, all following the same dizzy collision course into other strings of five.

Soon our lights disturb the moths. No, they're not attracted, they're repelled. Normally they'd be up there in flight for several hours. Now they go to bed early, scurrying on foot across the rock surfaces and down into the dark.

By torchlight we pack up and trudge down the mountainside to the vehicle, puffing moth dust from our lungs as we go, still feeling the awesome influence of something beyond our understanding and very ancient.

'They were probably doing all that when we were just a gleam in a dinosaur's eye,' says Jim.

But why? What's it all about? Why do these Bogong Moths come so far up the mountain tops, so far from home, on such a hazardous journey? How do we interpret the dance that takes them even higher, whirling against the sky without apparent purpose, dizzily zig-zagging, faster and faster? Is it mere exercise? Sheer joy of living? A ritual of communication? Perhaps we'll never know.

Perhaps it's simply what Shelley said it was: the desire of the moth for the star.

OPPOSITE: *In good seasons the moths are so numerous they spill out into the light.*

16 The spider that's got the drop on insects

One of my most memorable moments was the discovery of my first Net-casting Spider, in the garden one night many years ago. There she was, unmistakeable in the torchlight, a big, leggy spider hanging upside down in a bush, holding what looked like a miniature skein of white knitting yarn.

I'd never seen *Dinopis subrufa* before, but from illustrations in books she was immediately recognisable. And because I'd read so much about this spider I knew that what she held was an ingenious device for catching insects,which with patience I might be able to see in action.

A discovery doesn't have to be an 'original' one in the narrow sense of the word. A personal first sighting is no less novel because someone else has already got in first with a name or a description in a scientific journal. In fact it can be more exciting.

However, in the alien-seeming mini-world of the spiders it's useful to have a few introductions from those who have trodden the main tracks. But make no mistake, there are uncountable byways still waiting to be explored. The gaps in our knowledge can be closed by any carefully observant visitor from the mega-world.

But how, in my own garden where every nook and cranny and, as I thought, every inhabitant was familiar, had I missed seeing *Dinopis* before? Well, the focus up until then had been on insects. My 'spider phase' was just beginning. For me the book *Biology of Spiders* by English naturalist Theodore Savory had become riveting bedtime reading; it was taking me into a world that science fiction writers would find hard to dream up.

Anyhow, there was *Dinopis*, looking so much like a dead leaf caught in a bit of web that I wouldn't have seen her at all if my torch hadn't lit up the shining blue-white structure of her net. It lit up something else, too: a splodge of white faeces on the ground directly below her. I thought nothing of this at the time, but those droppings were there for a purpose.

Over the next few years I got to know this spider well enough to fill in some of the unknown facts of her life history and behaviour. Watching and taking notes night after night I was able to work out the sequence of net construction and other details of this remarkable spider's life history. I must explain here that when I refer to a spider as 'she ' I'm being neither politically correct nor sexist. To use 'it', except when speaking in general terms about a spider, is impersonal as well as vague; like us, spiders are separately either male or female. But mature male spiders don't make webs or live very long, so most of the spiders we see in webs around our garden are females.

Until the English language takes the sensible course of adopting a unisex pronoun to replace

ABOVE: *The netcasting spider* (Dinopis subrufa) *waits for an unwary insect to pass below her.*

RIGHT: *Two enormous eyes give this spider a keenness of vision unmatched among her kind.*

he/she, this niggly little problem will be with us. So unless I'm referring to a specifically male function I shall go on calling my spiders 'she'.

This is not to say that the male *Dinopis* is in any way a mere cipher. Up until his last moult he makes the same kind of net as the female. After that he never feeds again. His last and vital act before dying is to find a female and mate with her. But first he strings out a line to her scaffolding and

JIM FRAZIER

TOP: *Dinopis weaves her silk net, working head-up on a special scaffolding.*

CENTRE: *The spider works by touch alone, the net shaped by her claws as they draw out the silk threads.*

BOTTOM: *The finished net is so elastic it can stretch to many times its apparent size.*

courts her by patiently plucking on the thread of communication. After several hours she gets the message and comes to meet him.

Dinopis has close relatives in Papua New Guinea and in tropical America. In Australia several species are common in south-eastern forests and gardens. You won't find *Dinopis* easily, though. She spends the day close to the ground posing as a pair of crossed sticks, and if it weren't for her white net she'd be just as hard to find at night. But this is a spider worth going out and looking for. It's best done at night, shortly after dark when she's setting up her operation. Your torchlight won't worry her.

There are several things that set this remarkable spider apart from her fellows. Her eyes, for instance. Like most of her kind she has eight eyes. But two of them are enormously big and round. They're set in a fringe of ginger 'lashes', rather like the eyes of a friendly Hereford cow. There the resemblance ends. *Dinopis* is no gentle, myopic herbivore but a rapacious predator with a keenness of vision unmatched among her kind. Research has shown her eyes to be *12 times more sensitive to light than our human eyes*.

Most of the spiders we see around trap their victims passively in their webs. They're short-sighted, sensing their victims solely by vibrations of the web. But *Dinopis* is an active ambusher, relying on her acute night vision and hair-trigger reactions. With few exceptions she catches not flying insects, but insects that walk directly beneath her net.

You could find a parallel to this spider in those infra-red devices that open doors for you when you leave the supermarket with your arms full of parcels. Just when you expect to crash into them, the doors open to let you through. *Dinopis* gets her parcel afterwards, though. She is triggered into action when an insect passes directly beneath her. She's down there in a split second and her victim's in the bag.

But that's just the final coup. There's a lot of preparation first. Every night—sometimes several times in a night—the spider must make a new net. She weaves it with a complex kind of fuzzy silk that only a few spiders can produce.

Spider silk is a glandular secretion that in most spiders comes out only through the finger-like spinnerets at the tip of the body. *Dinopis* has the spinnerets but she also has an extra set of glands and an extra spinning organ. The silk she uses for her net comes not from the spinnerets but from a small perforated plate just forward of them called the *cribel-*

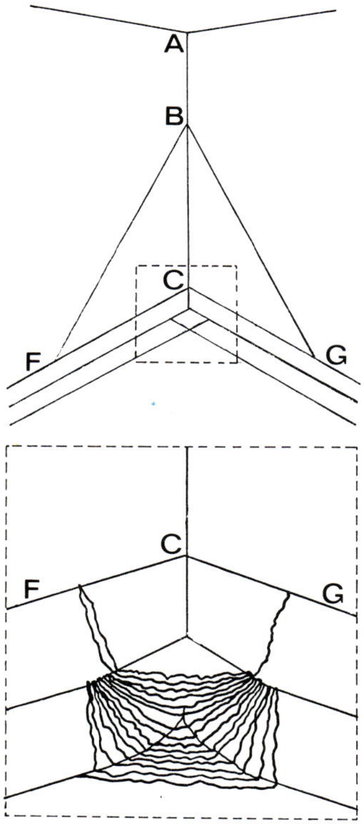

ABOVE: *Diagram of the basic framework erected by the netcasting spider. It is both the loom she weaves her net on, and the launching pad for this remarkable predator.*

lum, or 'little sieve'. To draw out this special silk the spider has an elegant little comb on each hind foot, shaped like one side of a feather. The silk comes out as two separate threads that join together.

The finished net is a fragile and beautiful artefact. It takes a long time to make and it's easy to see. Not so obvious but just as important is the larger structure of plain silk threads from which the spider operates. It's a combined watchtower, trapeze, launching platform and weaving frame.

Although the angles and relative lengths of the threads that make up this structure may vary according to the situation, the basic design can usually be seen. The threads of the structure are too fine to show up well in a photograph so I've borrowed some of the diagrams I drew for an article in the scientific journal *Australian Zoologist*.

Dinopis usually leaves her scaffolding structure in place during the day, hiding somewhere nearby. Soon after dark she emerges and suspends herself upright by her first and second pairs of legs, grasping the lines I've marked as AB, FC and CG. This is her net-making position. While she combs out the silk threads with her two hindmost legs, turn and turn about, she uses the claws of the third pair of legs to grasp the threads and weave them together.

The net starts out as a narrow horizontal ribbon, but as it grows the two ends gradually turn upwards. Finally the space in the centre is filled in, completing a rectangular sheet. Some minor framework threads are cut, some vertical threads are added, and that's that; but there's more to do yet.

The spider moves from a head-up position to a head-down one and cuts through the vertical thread BC as she goes. So why doesn' t the net collapse and drop down? It can't do that because *Dinopis*, hanging from her own safety line, has a firm hold of the framework. Next she moves down a bit and stretches out her long front legs to touch the ground or the leaf or stem below her. She's measuring the exact distance she'll have to drop from her ambush position.

Back on the scaffolding now, the spider grasps the corners of her net with her four front claws and tests it, flinging her feet apart. You can see the remarkable elasticity of the net as it stretches to several times its apparent size.

There's one more job to do. Before she bunches up the net and goes all quiet and watchful the spider flips her abdomen over and drops that splodge of white faeces I mentioned, directly underneath her.

The ambush site isn't chosen at random. The net may be built over a stick on the ground or a horizontal stem on a shrub. Sometimes the spider angles her structure against a vertical surface such as a tree trunk or fence where insects are likely to crawl.

Hanging over a busy insect highway like a police helicopter on a holiday weekend, *Dinopis* can make

The spider drops a splodge of faeces on the ground below—perhaps to act as a sightboard for her catch.

Her victim secured in a tangle of silk, the spider starts her meal.

several snatches on a good night. If she has no success over several nights she will move to a more promising spot. Sooner or later some wandering insect or other will cross her line of sight; it might be a beetle or a cockroach, or even a large ant. If that passing insect should happen to look up, its last sight would be a frightening one. Ape-like brows overhang the spider's enormous black eyes. The hairy chelicerae that bear the fangs are worked together as a stage villain might grind his teeth . . . *Dinopis* has another name, 'Ogre-faced Spider'.

But there's little time to take all this in once the spider spots her victim. She tenses, trembling a little on her silk support. Now—and if you blink you'll miss it—for a fraction of a second only she lets go her hold on the trapeze and then grabs it again. And in that split second she has dropped the measured distance to her prey. She flings her net wide, scoops up the insect, springs back with it on her trapeze. A few threads secure the victim, a bite despatches it and the spider settles down to her meal. Already her long legs are at work weaving a net to catch the next one.

And what of the white 'marker' the spider drops below her at the ambush site? Is it just that, a kind of sight-board to show up her passing victims? Is it a bait to bring scavengers to the scene? Or is the spider just making herself comfortable before she settles down?

I favour the first explanation—but your guess is as good as mine. As I said, there are lots of gaps in our knowledge just waiting for someone with patience and enthusiasm to come along and close them.

17

Dragons with cloaks and beards

In early Victorian times a lady setting out on a solitary walk was advised to take a large umbrella with her. Any dangerous male met along the way could then be put to flight by its sudden unfurling and a series of rapid lunges with the sharp end.

Quite disconcerting, I should think, especially if at the same time the lady bared her teeth and hissed. The strategy certainly works well for the Frillneck Lizards.

The Frillneck faces its enemies with jaws agape, the colourful frill held erect around its head by a set of fine, flexible supporting rods, like the spokes of an umbrella. And here's the fascinating part: it's the opening and closing of the jaws that automatically opens and closes the frill. The 'spokes' are extensions of the hyoid bones at the base of the tongue.

The lizard's threat display is no mere bluff; it will readily charge and chase an attacker. This is of great benefit to a photographer wishing to get a head-on view of the frill. You have to be quick, though, because the Frillneck is quite likely to close its jaws on the lens before you've had time to hit the button.

But Frillnecks are normally quite peaceable reptiles. They feed on insects and other small animals, foraging both on the ground and in the trees. They spend a lot of time just lying motion-

The Frillneck Lizard (Chlamydosaurus kingi) *faces its enemies with jaws agape and frill unfurled.*

Attacks usually come from predatory birds, so a Frillneck in the open must keep a wary eye on the sky.

less, on a log or up a tree. It's the hardest thing in the world to find a Frillneck when you want one, because camouflage is their first, and most effective, line of defence.

Caught out on the ground, the lizard will attack if you or a swooping bird are close enough to catch it. But sooner or later it will turn and head for the nearest tree. A Frillneck on the move is a sight worth seeing, and its extraordinary gait has given it another common name, the Bicycle Lizard. It runs with astonishing speed, erect on its muscular hind legs, front feet dangling and nose up in the air. All the time it moves its head from one side to the other, keeping an eye out for signs of pursuit.

In 1884, in a paper written for the Linnean Society of New South Wales, one C. W. de Vis first described this characteristic gait in print. He saw the lizard 'trotting out briskly on its hind legs, its fore-paws hanging down affectedly and its vertical line to the very snout stiffened at an angle of 60 degrees . . .' then stopping periodically to '. . . erect its frill and at the same time turn its head enquiringly from side to side'. Now, a hundred years later, I'd liken it to an anxious cyclist caught on Sydney Harbour Bridge in peak hour traffic.

But the display alone is usually enough to deter birds of prey and other enemies. The suddenly expanded frill says: I'm bigger than you think! And all of a sudden it seems too much of a spiky mouthful.

Speaking of mouthfuls, the Frillneck's scientific name *Chlamydosaurus kingi* might strike you as being one. But it's quite simple when you work it out. *Chlamydo* comes from the Greek word for garment or cloak, and *saurus* just means lizard. A cloaked lizard.

At rest the frill would be better described as a shoulder-length cape than a cloak. A baby Frillneck has little more than a fold of skin with a

RIGHT: *A baby Frillneck does its best to put on a threatening display towards a human finger.*

LEFT: *A young Frillneck has snapped up a fly at lightning speed.*

FAR LEFT: *Being peaceable lizards, their first line of defence is camouflage.*

scalloped border, relying mainly on camouflage for protection. In fact Frillnecks of all ages spend much of their time basking in the sun pretending to be broken branches or stumps. The frill, with its colours hidden, becomes nothing more than an aid to the camouflage.

The first description we have of a Frillneck comes from Allan Cunningham, who was here on a plant-hunting expedition for Kew Gardens in 1820. He called it 'a lizard of extraordinary appearance . . . with a curious crenated membrane like a ruff or tippet around its neck.'

There is, indeed, nothing quite like our Frillneck anywhere else in the world. It occurs only across the north of Australia and down the east coast to the vicinity of Brisbane, and it has no close relatives even in its native land. Most people know it best from television coverage, from its image on our old two-cent coin and on wildlife car stickers and posters and magazine illustrations. But in the south its name is very often mistakenly given to another kind of dragon-type lizard.

More times than I can remember I've had someone tell me about the 'Frillneck' they found in their backyard. It always turns out to be a Bearded Drgaon. The Eastern Bearded Dragon, *Pogona barbata* (both words mean bearded, the first in Greek, the second in Latin), lives in most parts of the southern half of the continent, and it's often seen in suburban gardens. Let me correct that. It's often in suburban gardens but seldom seen there.

Like the Frillneck, a Bearded Dragon can become almost invisible when it's at rest in a tree or on a stump or fence-post. It has a sneaky way of slithering around to the far side long before you spot it. The game's given away, if you happen to be looking in the right direction, when a suspicious eye slowly comes into view as the dragon leans out to check on you.

If you meet a Bearded Dragon face to face it will flatten its body, expanding it sideways, no doubt to make it look bigger. At the same time it will inflate the pouch under its chin, which is quite different from the Frillneck's winged flaps. The spiky scales stick out around the pouch making it even more like a bristly beard.

At the same time the dragon will open its jaws and hiss, displaying a bright orange-yellow mouth with a formidable array of little pointed teeth. Make no mistake: this lizard will readily snap its jaws shut on a human finger, but this

The common Bearded Dragon (Amphibolurus barbatus) *is often confused with the Frillneck, which only occurs in the north of Australia.*

In spite of their fearsome appearance, Bearded Dragons are harmless feeders on insects and flowers.

causes more shock than pain, and no Australian lizard of any kind is venemous.

Both the Frillneck and the Bearded Dragon belong to the great family of lizards grouped together as 'dragons' because of some general characteristics they all share. And back on the subject of scientific names, the one used for the dragon family, Agamidae, has nothing to do with dragons.

There's a little lizard of the Caribbean Islands of the Americas known in the local language as Agama. This name was apparently and rather surprisingly chosen by zoologists for a genus of lizards that doesn't occur in the Americas at all. In turn this *Agama* gave its name to the entire dragon family, with members on all but the American continent.

But one group of agamids does reflect the world of mythology in its scientific name.

The dragons of myth and, who knows, perhaps of race memory, were called by the ancients *draco*, which means 'one who sees clearly.' It seems an odd choice of name when you think of the many aspects of dragonhood that are far more striking than keen sight. However, it's from this word *draco* that we get our word dragon. *Draco* was the name given to a genus of 'flying lizards', the little arboreal dragons of South-East Asia that use wing-like membranes between front and hind legs to glide downwards. As mythical dragons were often depicted with wings, it's an appropriate name.

While present-day dragons are not very big and they don't breathe fire or carry off maidens, the emotion they inspire is more often fear than delight. But not always. A few years ago our Frillneck Lizard took off as a minor cult figure in Japan after it appeared in a televised car commercial. The commercial did a brilliant job of selling not the car but the dragons. They appeared as effigies in metal and plastic, as badges and brooches, and printed on posters, T-shirts and car stickers—'Love is a Frillneck Lizard'? Japan went wild over the wonder from Down Under. Then like all such popular crazes this one over-reached itself. Now I'm told every Japanese child wants a baby wombat for its birthday. Or is that already out of date?

But what was it about this particular lizard that could even for a short time turn the affections of a nation so readily away from the conventional cuddlies?

I think I know. I think it was the sight of that cute, umbrella-like frill in operation. After all, it was in the East that the principle of the umbrella or sunshade was invented more than 3000 years ago. Re-invented, I should say. Our Australian Frillneck got in first.

18 Killer caterpillar routs rural invader

Prickly pear—the green invader that laid waste to millions of hectares of Australia's rural land.

As a peripatetic naturalist I've found over the years that forced dalliance by the roadside on a long journey from A to Z can be both relaxing and fruitful.

I would say to the poet who wrote the lines:

What is this life if full of care
We have no time to stand and stare?

that a crouching or squatting position often gives you a much better vantage point than standing.

After all, who wants to stare at cows or sheep when there are bugs having fun on bushes, ants carrying things in and out of holes, spiders weaving, grasshoppers fiddling, stick insects challenging your perceptions, flies—well, we'll leave flies out of it. But there's always some small entertainment going on down there, perhaps only a single act in the continuing drama of life at the grass-roots level. Enough to occupy your mind for the required few minutes.

Next time you're travelling through warmer parts of the eastern states and need to make one of those stops do it, with care, beside one of those exotic Prickly Pear cactus plants you see growing at the roadside. How green and fat and healthy it looks at a distance! Close up, it's more than likely you'll see signs of distress. Some of the segments may have turned brown or collapsed in a shrivelled heap. Others, looking pale and withdrawn, may drop some unpleasant-looking goo out of a small hole.

If all seems well, run your eyes over the tufts of barked spikes on the succulent blades. See that bit of dried grass stem stuck on one of the spines? Well, it's not a grass stem at all. It's actually divided into tiny segments like beads strung end to end. You're looking at the deadly weapon of *Cactoblastis cactorum*, slayer of cactus. More prosaically, it's the egg stick of a small, nondescript moth that nevertheless saved an entire agricultural community from ruin, three-quarters of a century ago.

But the story of cactus and moth begins much earlier, and its link with the First Fleet, and with the red-coated soldiers of early settlement, is worth a brief mention.

En route to Australia in 1788 the First Fleet under Captain Phillip called at Rio de Janeiro. It seems that Phillip took on board there some of the little scale insects that produce cochineal, the

Eggstick of the American moth that saved the day (Cactoblastis cactorum).

ABOVE: *A mature* Cactoblastis *caterpillar.*

TOP: *The eggs hatch and the tiny caterpillars have their first meal before boring into the cactus.*

RIGHT: *Inside the cactus blade the caterpillars grow fat on its succulent flesh.*

red dye used in those days to colour soldiers' uniforms. He also took aboard some plants of *Opuntia monocantha*, the 'Tree Prickly Pear' that is the food plant of the Cochineal Insect.

As governor of the new Australian colony, Phillip apparently intended to start a home industry of cochineal production. I don't know what happened to the industry but a century and a half later the whole scheme backfired when the Tree Prickly Pear became an agricultural pest by forming impenetrable forests.

It was a later migrant, though, a smaller relative of the Tree Prickly Pear called *Opuntia inermis* that did a great deal more damage, and cost the country a fortune in personal distress, money and resources.

The common Prickly Pear seems to have arrived in eastern Australia about the middle of the last century. By that time cactus plants from the New World had been introduced as quaint exotics to many northern countries. Perhaps *O. inermis* came in as a treasured glasshouse potplant that would surely flourish in the open in such a favourable climate.

Flourish it did. In fact there was no holding it ... and I'm not referring to the spines. *Opuntia* got the scent of freedom and went exuberantly feral. With no competition and nothing to check it, and indeed helped later by half-hearted efforts to chop it out, the Prickly Pear spread across the land like a green tide.

When a plant migrates, its natural checks and controls are usually left behind. Australia, with no native cactus species of its own, had none of the cactus eaters (insects and mammals) that would normally keep such plants balanced with their enviroment. There was nothing here capable of taking on the Prickly Pear. We were on our own, out on a pretty uncomfortable sort of limb.

By the early decades of this century Prickly Pear had turned into a virtual wasteland something like 26 million hectares of land, ranging from Mackay in Queensland to the Hunter Valley of New South Wales. Farms were ruined, cattle starved, crops had no growing space; families fled as homesteads disappeared under thickets of prickles.

No use now the obvious form of attack: relentless slash-and-burn or chemicals. The enemy had won too much ground. Official estimates put the total cost of *initial clearing only* at $100 million, a fortune for those times. On low-yield grazing and farming land no farmer could afford the expense. They just walked out.

Soon after the First World War the Prickly Pear

Board was set up to get to grips with the problem. The biological control of pest species was a new thing then, but this was the route the scientists took. The world was searched for suitable insects to be introduced into affected areas. There was some success with the Tree Prickly Pear—a Cochineal Insect from India, released after trials, brought the big cactus under control and still keeps it that way. But the Common Prickly Pear, the real villain, proved to be a tough one.

Biological control experiments take time and care. It's essential to test an introduced species over and over before setting it free, making sure it can't do more harm than good in its new home. From America a selection of 50 different species known to attack *O. inermis* was brought to the Board's Brisbane laboratories for testing. Some, passing safety and survival tests, were bred for release but none proved wholly effective against the enemy.

In 1925 a ship arrived from the Argentine carrying several thousand caterpillars of a small moth, *Cactoblastis cactorum*. What followed turned out to be the greatest success story of biological control ever told, and it remains so.

Once again the scientists did their testing and their breeding, producing millions of Cactoblastis eggs for distribution. The female moth lays her eggs in the form of a stick attached to a cactus spine. Out of the eggs come handsome little orange caterpillars banded with black spots that bore their way into the cactus blade and feed on the succulent flesh inside. As they feed, their droppings pass hygienically outside as a succulent mush—sure evidence that the plant is under attack. The fully fed caterpillars leave and make their cocoons on the ground. The cactus, now a shrivelled ghost of itself, topples to the ground.

So out went the egg sticks to selected testing areas, to be stuck on cactus spines. The effect was immediate and dramatic. New breeding stations were set up all over the affected countryside. And already the moths were breeding naturally, accelerating the effect. By 1933 they had decimated the cactus hordes and cleared the land again for crops and pastures. The battle was won. The people returned to the land.

Cactoblastis cactorum, the moth, is now a fully naturalised Australian. *Opuntia inermis*, the Prickly Pear, is now outlawed—oh, it's still around, we've all seen it, and quite a pretty addition it is to the Australian landscape. But wherever it tries to rear its spiky head too high it's cut down to size by

ABOVE: *While the cactus succumbs and dies, the caterpillars turn into moths, ready to lay their eggs on another plant.*

LEFT: *The Cactoblastis Memorial Hall near Chinchilla, Queensland—tribute to the insect that saved a rural industry.*

a new generation of cactus-slaying caterpillars.

On a dusty outback Queensland roadside, not far from the town of Chinchilla, you might one day in your travels pass a small wooden building with an iron roof. Hens will be scratching around it in the bare earth. Ducks will be paddling lazily on a muddy pond nearby. You may not see any people at all, but a sticker or two advertising perhaps a fair or a tennis tournament, or just the next meeting of the Country Women's Association, hint at a lively and thriving rural comunity going about its business.

However, what gives this modest building its distinction is the sign blazoned grandly across the front under the inverted V of the iron roof, proclaiming it as the CACTOBLASTIS MEMORIAL HALL. It's probably the world's only memorial to a moth.

19 A rare and comely insect from Arnhem Land

A friend calls it the Paradise Grasshopper. Some call it Leichhardt's Grasshopper. There's no widely accepted common name for it, which isn't surprising. This beautiful insect is found only around rock outcrops in remote, largely unexplored parts of Australia's Northern Territory.

So we're stuck with *Petasida ephippigera*, and it's almost as hard to find a meaning for this unwieldy scientific name as it is to find the grasshopper itself. Why bother? you might ask. Well, I bother because I find the natural history of words fascinating, and tracking down the meaning and relevance, or lack of it, of scientific names can be surprising fun.

But this is a hard one. *Petasida*—well, the ancient Greeks used to wear a wide-brimmed, conical-crowned hat called a *petasos* but I can't see the connection. There's a family of jellyfish called Petasidae. I suppose a jellyfish, being round, could at a pinch look like a *petasos*. But a grasshopper? Wrong shape entirely.

As for *ephippigera*, all I've discovered is that the word could have something to do with (1) being an old horse or (2) carrying an old saddle. I leave it to some classical scholar to enlighten me. It's not possible to ask the man who would know best because he gave the grasshopper its strange name back in 1845.

Not very much is known about the natural history of this marvellously colourful grasshopper, but the history of its discovery is fairly well documented, and it's linked with some famous names from our early days.

The first specimen of *P. ephippigera* was collected by the purser of the ship HMS *Beagle*. In 1839 as part of a survey of the north Australian coast, the ship sailed 260 kilometres up the Victoria River. Judging from later knowledge of the kind of terrain the insects like, it's been suggested that this is probably where that first specimen was found.

Ludwig Leichhardt, the explorer, caught the next one in 1845. Leichhardt was close to the headwaters of Deaf Adder Creek at the time, right up near the top of the Arnhem Land Escarpment.

RIGHT: *A Paradise Grasshopper photographed by courtesy of the CSIRO, Canberra, shortly after its rediscovery.*

JIM FRAZIER

ABOVE: *Confined to limited habitats in Arnhem Land and the Katherine River Gorge, this rare insect is seldom seen.*

TOP: *A juvenile Paradise Grasshopper on its food plant is not easily found in spite of its bright colours.*

RIGHT: *The Paradise Grasshopper* (Petasida ephippigera) *was discovered in 1839 and later lost to sight for more than a century.*

His diary records that there were 'a great number' of them about. A third specimen was collected by J. R. Elsey, surgeon to Gregory's North Australian Exploring Expedition of 1855–6.

After that, for more than a century nothing was seen or heard of the grasshopper. Then in 1971 Dr. J. H. Calaby of the Commonwealth Scientific and Industrial Research Organisation (CSIRO) in Canberra found one immature male on a trip to Arnhem Land. It was at Mt. Brockman, in the South Alligator River area.

There must have been a sudden population explosion about that time. Shortly afterwards some Aboriginal people collected one adult and 15 nymphs at Maningrida on the coast. The odd thing is that the insects were so unfamiliar to these local men that they had to take them to the Maningrida Progress Association to find out what they were!

Now several other finds followed rapidly and the insects could be linked at last to some of the plants of the region. Their food plants turned out to be aromatic species of *Pityrodia* and *Dampiera*.

In the meantime scientists at the CSIRO Division of Entomology in Canberra had successfully brought some of those 15 nymphs through to adulthood, feeding them on plants donated by the nearby National Botanic Gardens. It was there at the CSIRO that I first saw these striking insects, and by courtesy of Dr. Ken Key was able to photograph them during a brief visit in February 1971—on the front lawn of the CSIRO with a gale blowing. Ten years later I saw and photographed them in the wild at Kakadu National Park, and my last encounter with these spectacular grasshoppers was on a wild and rocky hilltop overlooking Katherine Gorge.

Incidentally, early in 1972 a mature male *P. ephippigera* was found by a soil contractor 11 kilometres from Canberra. Still quite a long way from home, but hopping well . . .

JIM FRAZIER

The Dingo (Canis dingo)—*has it an undeserved reputation for cowardice and treachery?*

20

The Dingo – scapegrace or scapedog?

Give a dog a bad name? Well, the Aborigines gave the Dingo its name. We only gave it its reputation.

Drought-killed cattle help outback Dingoes to survive hard times. It's this carrion-feeding habit that links them in people's minds with jackals and coyotes and pariah dogs: cringing, cowardly and altogether disgusting. (Never mind what Rover was eating on the footpath the other day, that has nothing whatever to do with it . . .)

Is it possible that these lowly-esteemed animals have become scapedogs for some of our nastier human habits?

Predator and carnivore it certainly is, but there's nothing to justify the Dingo's long-held reputation as a 'treacherous coward'. Did Dingoes steal from the white men's camps in the early days of Australian settlement? Hardly a matter of treachery; we brought the food into their territory. Did they kill and eat helpless sheep and cattle? Well, let's not cast that particular stone—we do it ourselves, perhaps less mercifully and certainly on a larger scale.

You'd hardly think from the Dingo's reputation that it is the same animal we call Man's Best Friend. The same old *Canis familiaris,* domestic dog, guardian of our homes and our lives. So long domesticated, in fact, that the identity of its wild ancestors can only be guessed at.

Yes, the Dingo is that dog, too. A rather broader skull and bigger teeth, some variations in behaviour. Just enough difference to justify the addition

Relative newcomers to Australia, Dingoes probably arrived here with the first human migrants.

JIM FRAZIER

JIM FRAZIER

JIM FRAZIER

Dingoes are now bred in captivity for wildlife zoos and as family pets.

of a third, distinguishing name. Where the domestic dog's full scientific name is Canis familiaris familiaris ('very domestic dog'), the Dingo is *Canis familiaris dingo.*

It's thought the Dingo came to Australia with the Aboriginal people many thousands of years ago. Perhaps it was brought in as an already established hunting dog. Perhaps there was a closer bond. Anyway, Dingoes were certainly among the first wild animals to share Australian man's camp fires, but where and when the association actually started is less certain. Some say there's an ancient Indian connection.

However we classify it, like all feral carnivores the Dingo has been bad news in the past for some native marsupials. It may have been the Dingo that drove the Thylacine to extinction on the mainland. There are no Dingoes in Tasmania, which was (perhaps is) the stronghold of the Thylacines.

The Dingo has been a strong suspect in the disappearance of at least one human infant. Notable is the tragic story of the Chamberlain baby whose mother, Lindy, found guilty of her baby daughter's murder, spent many years in prison until totally exonerated. It was a Dingo that took the child from its parents' tent during a camping trip to the outback.

The Dingo's effect on the grazing industry is still a matter of debate and research. Its proven diet is largely rabbits and small native lizards and insects. Where these are scarce, Dingoes will hunt as a group to bring down kangaroos, wallabies and wombats.

The first record of a Dingo sighting we know of was by a Dutchman, Jan Carstenzoon. He took a party ashore in Queensland in May 1723 and saw footprints of dogs, and later 'a great number of dogs'. Later in the same century other Dutch explorers mentioned dogs with yellow colouring and with long ears.

With the early English settlers came the first proper descriptions, and the first character blackening. In 1802 one George Barrington describes the Dingo's 'ill-nature and viciousness'. A little later, in 1834, George Bennett writes of the

Dingo's ability to withstand pain, and, in the same sentence, its cunning.

Bennett then documents two incidents which quite put the boot on the other foot, the human one.

In the first, a Dingo is beaten so severely that all its bones are thought to be broken. What a surprise when this animal, thought to be dead, gets up, shakes itself, and heads for the scrub! Cunning?

Worse follows. A Dingo, supposed to be dead, is brought to a hut for skinning. Someone notices that its lip is still quivering. Nevertheless, skinning commences. The man leaves the hut to sharpen the knife. What a surprise when he comes back to find the animal sitting up with its flayed skin hanging over its face! Vicious? Ill-natured? A saint would be after that kind of treatment.

Dingoes have been bred in captivity now for some time. They've even been kept as pets and people with close knowledge of them are devoted in their defence. However, it should be remembered that the owners of several vicious breeds of domestic dog, known for fatal attacks on children in particular, are equally devoted. This is not to cast all canines in the same mould, any more than we do with humankind.

There's a feeling in the air that the tide of public opinion against the Dingo has been turning in recent decades. There's a certain protectiveness. A new empathy. Our handsome wild dog is, I think, about to have its day.

Dingoes can and do mate freely with domestic dogs. It's a possibility that the true-bred *Canis familaris dingo* may soon disappear as a wild animal and remain only as a curiosity in zoos. Or as a gleam in the eyes of our well-fed, well-loved, four-footed best friends.

Alert and wary, a wild dog is always on the lookout for danger.

21 Moths in the Movies

JIM FRAZIER

ABOVE LEFT: *In due time a spiky black and white Hercules caterpillar hatches.*

ABOVE RIGHT: *The skin of the moulting caterpillar peels off backwards.*

Regularly every year, deep in Queensland's tropical rainforest, a time clock ticks into its final phase in the making of the world's biggest moth. Biggest, that is, if it's a female that comes bursting out of the cocoon. The 5-inch long, blue-green caterpillar that gave up eating three months earlier to pupate in the canopy would have given no hint of its sex.

Coscinocera hercules, the Hercules Moth, lives only in tropical Australia and Papua New Guinea. Based on wing area, the female beats the Atlas Moth (Atlas atticus of South-East Asia) by a short margin for the title of 'biggest'. Many birds are smaller. But with her furry body already distended with eggs when she emerges from the cocoon, the female leaves most of the flying to the male. He, smaller but distinguished by elegantly tapered hind wings, will use wings and antennae to find his mate, often travelling long distances.

The huge, doubly feathered antennae of a male Hercules are his scent receptors. They're the insect equivalent of our noses but they're many times keener. They need only collect a few molecules of the female's sex pheromone from the air to lead the male through the crowded rainforest to where his mate sits waiting in the dark.

The pair will mate, and from the female's eggs in due time will come tiny, spiky white caterpillars. They give no hint of the colourful giants they will become, after a season of non-stop munching on lush rainforest leaves.

Caterpillars have always fascinated me, from the time I first watched a looper looping and poked a woolly bear to see it curl up. The more I learned about the character and habits of moth larvae the more the fascination grew. I can't remember ever having my windows clear of the silken ladders of case moths, or my kitchen benches free of the litter from gumleaf-chewing Chinese Junks, mad hatterpillars and others.

But one year my caterpillar house-guests included some that were more exotic and certainly much bigger than anything found around our southern gardens and bushland. That was the year my colleague Jim Frazier and I were making our film Desire of the Moth for the BBC's *Natural World* series.

Most of the subjects for the film had been drawn from the temperate bushland around my Sydney home. We had already been away on location, to the Snowy Mountains for Bogong moths, and southern Queensland for Cactoblastis, the moth that saved the pastoral industry from Prickly Pear. Now, for a starring role that would take in the whole of their life history, the Hercules Moths had come down from tropical Queensland to our studio. Not by their own wing power, but as eggs, and by airmail.

Writing a natural history film script is one thing; finding the cast is quite another. There was no certainty about obtaining the Hercules Moths. Even in normal seasons they're rarely seen, and our scouts in north Queensland said the weather was against us that year, with an erratic rainy season.

So we cheered when eventually a batch of eggs arrived from the tropical north. Eggs were the essential starting point if the entire life cycle was to be filmed. To rear a large tropical species in Sydney's temperate climate might seem a daunting task, but one thing was in our favour. Although the moth itself is restricted to the tropics, one of its larval food trees extends south

LEFT: *Tiny caterpillars give no hint of the colourful giants they will become.*

ABOVE: *An insatiable eight-centimetre long monster caterpillar for a house-guest.*

into the wetter parts of Sydney's bushland.

The small evergreen tree *Omalanthus populifolius*, or Native Bleeding-Heart, is regarded by some as a weed, but its heart-shaped leaves, ageing from green to scarlet, make it an attractive garden tree by any standard. And this is the tree the Hercules caterpillars thrive on.

Apart from a few casualties the spiky white hatchlings from the Hercules eggs did very well on a small potted tree in our film studio, supplemented with fresh branches from suburban gullies. But after a few weeks we found ourselves faced with a lot of insatiable, 8-centimetre long blue-green monsters—and a sudden dearth of *Omalanthus* leaves. It was a year of drought, and many of the moisture-loving young trees had died.

Jim searched the gullies far and wide while I phoned native plant nurseries and friends with bush gardens. We met the demand, but only just.

There were certain hazards to keeping the larvae on branches in bottles. On a morning inspection round I found one of them floating head-down in the water. What to do with a half-drowned caterpillar? Take it into the warm sunshine, mop it carefully

Inside its leafy cocoon a Hercules caterpillar prepares to become a moth.

Male Hercules Moth (Coscinocera hercules), *a giant rainforest species from tropical Queensland.*

The newly emerged female expands her crumpled wings.

with a paper tissue and apply mouth-to-spiracle respiration, or course! With all those breathing holes along its sides, giving the kiss of life to a giant, unconscious caterpillar is rather like playing a rubber mouth organ. But it worked. Or I like to think so.

At 10 centimetres long the caterpillars pupated, hanging from the branches in their cocoons of leaves and silk. We placed them for the winter in a plastic chamber, temperature and humidity controlled.

So far, filming the various stages of development had not been difficult. The emergence of a moth from the cocoon, though, would not be easy to catch and film. Other related moths, such as the Emperor, give you plenty of warning and time to set up a camera. They scratch away for hours inside their rock-hard cocoons, with a sound like a mouse gnawing wood. But their big cousin, the Hercules larva, spins a flimsier cocoon.

So Jim rigged up an ingenious early warning system. It was a tiny microphone fixed to each cocoon, leading to a loudspeaker in my bedroom. At the sound of scraping I was to leap out of bed, check the cocoons to find out which was active, phone Jim and keep the moth from emerging until he arrived.

The system worked so well we had to abandon

ABOVE: *The female Hercules Moth is the biggest moth in the world.*

it. The loudspeaker crackled every time a pupa turned over in its sleep, which, it seemed to me, happened many times throughout every night. With hindsight, though, I suspect it was just the loudspeaker getting its wires crossed.

But eventually we had a fine male and female safely out of their cocoons and eyeing each other across the studio. The male didn't wait long. The female laid her eggs. The last roll of film went through the camera. Our Hercules moths were now, so to speak, in the can, and we took a break.

LEFT: *The male's huge antennae are scent receptors used to find the female.*

22

Stand Aside for the Caterpillar Crocodile

If you've got an acacia tree in your garden you might have noticed roundish, creamy-brown growths appearing suddenly on the lower trunk, like some kind of furry mould, in early summer. If you were tempted to scrape them off you'll have missed out on following a fascinating life story.

Far from being a fungus or any other kind of growth, these fluffy blobs are the artefacts of an insect. Your tree has been selected as a nursery site by a small brown moth called *Ochrogaster contraria*, best known for the processionary habits of its caterpillars.

When the moth lays her eggs, masses of loose, hair-like scales come off the end of her body to cover and protect them. Watching an egg-laying female at work you'd wonder how all that fluffy stuff could have been packed so tightly around her. She may lay several batches of eggs, and other moths often lay on the same tree or an adjacent one. Seeing these excrescences growing in number, you're convinced the fungal infection is spreading and steps must be urgently taken.

The females of Ochrogaster contraria *cover their eggs with soft scales from their own bodies.*

But stay your hand or your step-taking foot. Keep a daily watch and you'll see how the blobs fluff out even more as the tiny caterpillars hatch from hidden eggs and mill around inside. After a few days they'll come out for their first meal.

Early November last year some egg masses were laid on an *Acacia pycnantha* outside my front gate. One morning a few weeks later when I went to fetch the milk there was a thread of small hatchlings moving head to tail up the trunk and along a branch. A quick glance, and it might have been just a fine crack in the bark, but a close look showed the leader inching (or millimetring) along in front. As the vanguard approached the twig tips, the rear echelons still poured out of their nurseries in ranks of four or five, thinning to single file higher up the trunk.

How to count such enormous numbers of 3-millimetre-long caterpillars when the eye could hardly separate them? Easy. I took the milk inside and got out my tape measure. I measured the entire procession from base to twig, added the shorter lengths of the additional flanking lines, and divided the result by one caterpillar length. Give or take a few, there were 2743 individual caterpillars.

Later in the morning the caterpillars were clustered at their feeding stations side by side around the rims of the leaf-like phyllodes. On acacias with true feathery leaves the little caterpillars scatter apart to tackle individual leaflets, and they're much harder to see.

For most of their lives processionary caterpillars are nocturnal, but for several weeks these went up the tree at first light, returning at midday to cluster on and under the leaf litter around the tree. Each caterpillar in its comings and goings left a thread of silk behind it. At first the threads were

ABOVE: *The patches of fluffy scales seem to expand as the caterpillars inside hatch.*

RIGHT: *The baby 'processionary caterpillars' move up a tree trunk for their first meal.*

visible only as a sheen on the bark. Reinforced twice a day this became first a narrow white track which the leader always followed exactly, then later as the caterpillars grew it was a shining highway up the trunk with access roads to branches and twigs.

Now the caterpillars fed only at night. On warm summer mornings when I went for the milk the trail end of a broad, furry phalanx was moving down instead of up, and milling around the entrance of a white silk tent at the base of the tree. In this tent the caterpillars spent their days and periodically moulted, adding spiky cast skins to the growing mass of dung pellets on the floor.

Protein-rich, leaf-browsing caterpillars are the insect equivalent of protein-rich, grass-grazing sheep and cattle. They are the main course on the menu of most insectivorous birds, lizards and other insects. This means that moths, in order to keep their species more or less in equilibrium, must produce vast numbers of eggs. If all those eggs hatched, if all those hatchlings survived to turn into more egg-laying moths, forests would fall. Life on land would come to a stop or at least pause for a breather.

Closer to home, my acacia tree this summer could not have survived 2000 unchecked, chomping mandibles. But by the time they were half-grown the caterpillars had been reduced to fewer than three hundred. I found a finger-sized hole through the tent wall one morning, and some disturbance inside. Perhaps a lizard or a currawong. Perhaps a boy with a stick.

At 35 millimetres long the caterpillars were ready to pupate. Would they burrow at the base of the tree? Not a wise move when you think about it. The moths that emerged next season would be likely to lay their eggs on the same food plant, so that the tree would have no fallow season in which to recover. The caterpillars would starve. Call it the wisdom of nature that sends them instead on an overland trek to a different site.

So I wasn't surprised, with all the comings and goings down my long driveway, to find evidence of their going one day recently—car-squashed furry bodies everywhere. Seeing the slaughter, a friend painted a sign, which she erected beside my drive. It reads: CAUTION! 5 KM. CATERPILLARS CROSSING. It's too late for this year's batch, but some of them will probably have made it successfully.

Driving out in the country you might have noticed the silk nests of bag-shelter moths strung up in roadside acacias. Some are as big as your head. The abandoned ones are often stained brown by the dissolved droppings after rain. The bag-shelterers build their roundish nests high off the ground, whereas the tent-shelterers put their tents around the bases of trees. It's curious that both these moths are given the same name and regarded as the same species when their behaviour is so different. But then it's only recently that zoologists have taken a big step into the present and started using behaviour to separate species.

Wherever they live, processionary caterpillars of any species are strictly NOT TO BE HANDLED. The fine, furry coats, so long and delicate when seen against the light, contain dangerous, urticating hairs. The word *urticate* is related to the scientific name of the stinging nettle, *Urtica*. It comes from the Latin *urere,* 'to burn'.

If you've ever brushed against a stinging nettle I can assure you that the effect of processionary caterpillar hairs is far, far worse, and burning is just the least of it. The hairs break off in bits and float about on to your face, down your neck, up your sleeves. Your eyes stream. Your face swells. There's intense itching that lasts for days. You may feel nauseated. You will look awful.

So why, you may ask, do I let these dangerous animals live in my garden, let alone suggest you do the same? Well, like so many of the animals we fear, they're only dangerous in their own defence. I was attacked because I was handling the caterpillars for several hours on a film set. Leave them alone, and they leave you alone.

After all, a garden is the stage for drama and comedy and the processionary caterpillars are part of the cast. If you stay in your priceless front row seat and watch the play to the end without throwing brickbats they'll do you no harm and at the end of the season move quietly away into the wings.

OPPOSITE: *The tiny caterpillars feed side by side on the acacia leaves.*

ABOVE: *Touch at your peril! Those attractive fur coats are made of dangerous urticating hairs.*

LEFT: *A silk tent at the base of the food tree makes a secure daytime hiding place.*

23

Autumn's Adagio for wings

When you're not looking out for them, seasonal changes steal up on you. Like age in the mirror when youth's still in the heart. Like the evensong of insects. Autumn has made subtle changes there. The summer singers have stolen away and less strident choristers have taken their place.

Instrumentalist of cool evenings and overcast days, a mole cricket (Gryllotalpa australis).

Is it the shortening of the days that triggers the autumn singers? Some of them start in January when night is only just starting its race to overtake day. Summer choristers like cicadas perform right over the crown of the year. Those noisy cicadas have just had an exceptionally good season; I heard Double Drummers still calling well into March, and after dusk at that.

Then one rather chilly evening I thought to myself, those cicadas sound different! Not quite the Double Drummer baritone, not quite the Green Monday tenor. More a kind of contralto purr. Restful. And rather melancholy.

Autumn had brought in the mole crickets, singers of dusk and cloudy days. The cicadas had sounded their last squawk.

Those of us with some of our roots still in England sigh occasionally for the strong seasonal rhythms of the northern year. The cycle from green to gold, from autumn colour to the winter starkness that makes spring a recurring miracle. Our evergreen antipodean year holds little inspiration for poets, prophets or philosophers. Just the same, there's a lot to be said for a low seasonal profile.

The English poet Keats says in his ode *To Autumn*:

> *Where are the songs of Spring? Ay, where are they?*
> *Think not of them, thou hast thy music too.*

and then he goes on to list the musicians of autumn.

In Keats' English autumn small gnats mourn and full-grown lambs (full-grown *lambs*—well that's what he calls them) bleat, Hedge Crickets sing, a Redbreast whistles and Swallows twitter.

Yes . . . and then? That's all? Well, I think we can do better than that down under. Actually, I think Keats could have done better, too. Perhaps he just wasn't very observant. But then, he was stuck with the shape of his ode and running out of lines. There's a lot to be said for free verse.

Prose is even freer, so let's see what sort of list I can make for autumn nights and days around the suburbs of our largest Australian city. Singers first. Those mole crickets, of course—two kinds.

The first of the mole crickets comes in as a throbbing continuo behind the late bird calls, and ceases shortly after dark.

The second kind does some nervous tuning up before settling into the rhythm of its repeated single call; this is the purring one. Like the cicadas, one mole cricket sets the pace. Its fellows follow from hidden places near and far.

Very hidden. Those underground instrumentalists are hard to track down. It's a funny thing about insects, they're always closer to you than they sound. You walk towards a cricket that's call-

ABOVE: *A field cricket raises a tune with his wings. Male crickets make music to establish territories and attract females.*

LEFT: *Not your usual insect musician, but a moth that uses a file and scraper on its wings to whistle up a mate* (Hecathesia fenestrata).

Freed from parental duties, this currawong will join others for autumn gossip sessions (Strepera graculina).

ing from way ahead—next thing you know it's behind you! It's different with frogs; they always sound as if they're actually out of reach on the far side. These are two things you need to know if you intend to spend much of your life on hands and knees checking on nature.

With crickets the thing to do is circle the general area with one ear to the ground—or as near as you can get it *and* circle at the same time. You'll know when you've got the sound fine tuned. The intensity of sound from a mole-cricket will astonish you and make your ears tingle and throb to an almost painful degree. And it's all done with a file and scraper on the bases of the wings, like a stick dragged along a paling fence.

Mole crickets call by day, too, on those fresh, drizzly days when the sandy bush soil gives off its special smell and the rock ledges run silver and the gumtrees stand stock still.

Field crickets are starting up now. They'll be trilling around our lawns right into winter, sweet and companionable outside bedroom windows on long, lonely nights.

You can get field crickets to the surface by running a hose into their burrows. If you must, do it gently. Keep one in a container so you can watch how it raises its front wings to scrape a tune with them. Only male crickets do it, and you might need to catch a female as an incentive, because this is a courtship serenade. But don't keep them captive too long or it will all be wasted.

Familiar birds that see the year through with us don't all stop their singing in autumn. Some of them just change their repertoire. The cool-season call of the Golden Whistler comes up from my bush gully lively and confident, like a greeting to a familiar friend. And the big black and white currawongs, now their families are off their hands, gather for guttural gossip sessions punctuated by squeals of 'Well I never!' The currawongs make another call now, in flight: a phrase whistled in a minor key. For this sweet utterance from such an unlikely beak I can forgive those rather sinister currawongs anything.

Then there are the strange little whistling moths that whizz dizzily at dusk above my front lawn. The males have a file and scraper on their wings, like the crickets, producing a persistent 'zizz, zizz, zizz', as they fly.

Musicians aside, there's lots of activity to be seen. Some hidden things become surprisingly obvious in autumn. The sun was just clearing the mist away one morning when my neighbour called me to the high fence surrounding her tennis court. The wire netting was embroidered all over with patterns of silk and dew, the work of hundreds of small spiders. Of course, the spiders and their webs were there before. We saw them for the first time only by courtesy of mist and sun rising together.

You don't expect the launching of new life in autumn. But this week a mass of Praying Mantises erupted from an egg case placed on my kitchen windowsill months ago. They skittered around, falling over their own spindly legs until I gathered them up and put them outside to take their chance. A few might survive. It takes only one raindrop to drown such a midget. It takes only one small wren or silver-eye to snap up 50 of them.

TOP: *The caterpillar of a Tussock Moth (Lymantriidae) will over-winter in a snug cocoon.*

Around Sydney, autumn is more evident in the suburbs than in the bush because of deciduous garden trees. But although the days do grow colder and darker there's no real need for foreign trees in parks and gardens to put up their shutters. I think they just colour up and drop their leaves to keep an ancestral memory alive.

As for the gum trees, they're not about to close down their chemical factories. They had their Fall in the full heat of summer when loss of leaves could be borne best. (And raking of their leaves by humans endured least.) So there are leaves a-plenty on these tolerant trees for the insects that feed on them.

The gorgeous long-haired caterpillar of the Tussock Moth is into its final stages, chomping away at its last long meal before over-wintering as a dormant pupa. The caterpillar-like larvae of the Gumtree Sawfly are only half grown. They feed communally. If separated they find their way together again by tapping their tails on the stem. Disturb them and they'll spit yellow globules of eucalyptus oil, supposed to nauseate attackers but harmless to humans. I like the smell. Strangely enough, these quite desperately gregarious larvae turn into solitary winged adults.

ABOVE: *Few of these newly Praying Mantises will survive the rigours of winter.*

So the list grows, and shows how life goes on in spite of the darkening days and the chilly nights and the lengthening shadows of a receding sun. And around Sydney's bushland you really don't have to look very far at all to see flowers-in-waiting already flushed with pink and yellow.

After all, it's no more than a hop, a skip and a jump now, out of the mists of autumn, across the shortest day, and into spring.

24 Birds that bind a chirming spell

A 'charm of finches'; what a delightful expression that is! One of those time-honoured collective terms, together with 'a gaggle of geese', 'a pride of lions' or, most felicitous of them all, 'an exaltation of larks'.

And how appropriate, you might be thinking if you're at all familiar with finches, either wild or caged. Well, it is appropriate, but not for the reason you'd expect, because the word charm as used in that phrase has a meaning different from the obvious one. It comes from an old English word chirm, meaning a twittering, warbling or humming sound.

Trace the word's ancestry back a bit further and you find the latin *carmen,* and for that word the English derivations include 'a song or tune', 'poetry', 'a prediction' and 'an incantation'. And it's from that last meaning that our modern word *charm* takes its usage. An incantation is a spell. A charm is a spell, so charming means spell-binding.

Red-browed finches (Aegintha temporalis) *come to a garden every day for a meal of millet.*

JIM FRAZIER

We can wear a charm against the nastier sorts of spell. But one who casts the nicer kind of spell is a charmer. The finches that delight us with their twittering are charming us with their charming.

To be honest, they're not brilliant songsters, and it's more their plumage and personalities that endear them to us. We're lucky in Australia that wherever we live there are finches of one kind or another to be seen, even in suburban gardens.

My sitting-room windows open on to a small paved area with a flowering shrub beside it. Every morning, you might say by simply waving my right hand, I make this little bush perform a magic trick. It starts with a quiver of leaves here, a quake there. A high-pitched squeak, then another, building up to a chorus of tweets and twitters.

Now all the branches shake their leaves together and the shrub comes alive as up to 50 small brown birds explode out of it onto their food tray: a charm of finches, come for their daily hand-out of millet seed.

If there's any movement indoors, if my cat's warning bell jingles, if a kookaburra zooms down out of a tree, there's an instant flurry of wings and all the birds are safely in the bush again. But only for a moment or two. The bravest hops down, the rest follow.

Did I call them *brown* birds? A closer look shows the green tinge of the wing feathers, the shiny bright red of beak, 'eyebrow' and rump. They're called Red-browed Finches and they're common all along the east coast and west into South Australia.

Australia has 21 different kinds of native finch, known collectively as grass finches to distinguish them from the related African weaver finches. While our finches make similar domed nests, they don't do any weaving.

Finches feed on seed, mainly of grasses, and on insects, especially termites. As seed-eaters need a lot of water, most finches live close to it on grasslands and savannahs, and along water-courses. The Red-brows in my garden pay regular visits during feeding to a pan of water nearby.

Some of the finches have specialised habitats. Painted Finches of the central and western deserts nest in prickly spinifex (*Triodia* sp.) and feed almost

OPPOSITE TOP: *The colourful plumage of a male Black-faced Gouldian Finch* (Erythrura gouldiae) *dazzles the eye.*

OPPOSITE BOTTOM: *The Red-faced Gouldian, a variant of the black-faced form.*

RIGHT: *In savannah lands of northern Australia lives the elegant Long-tailed Finch* (Poephila acuticauda).

JIM FRAZIER

RIGHT TOP: *Bird of rainforest and mangrove swamp, the Blue-faced Finch* (Erythrura trichroa) *is a recent migrant to northern Australia.*

RIGHT BOTTOM: *The Chestnut-breasted Finch of northern and east coast reed beds and river margins* (Lonchura castaneothorax).

FAR RIGHT BOTTOM: *The Crimson or Blood Finch of the tropical north* (Neochmia phaeton).

JIM FRAZIER

JIM FRAZIER

JIM FRAZIER

JIM FRAZIER

entirely on its seeds. Blue-faced Parrot-finches live on rainforest borders in North Queensland.

All the grass finches are attractive in their different ways. Some are strikingly patterned with spots, stripes and bars, such as the Zebra Finch—the widest spread and best known—and the Double-barred Finch of the north and east. Others, like the Crimson Finch, Star Finch and Blue-faced Parrot-finch have eye-catching colours. Long-tailed Finches with finely tapered black tails are the least typical and the most elegant, though not colourful.

There's one among the finches, though, that can match our gaudiest parrots for colour: the Gouldian Finch of tropical savannah lands. Some see it as the most beautiful of all our birds; and it could be a contender for world honours in the small bird class.

Your first encounter with a male Gouldian Finch is likely to leave you rubbing your eyes. Was it real? Did you imagine all that dazzlement of colour? Purple and yellow; cobalt, turquoise and green; jet black face and tail; even a touch of white to set it all off . . .

Although it's the Black-faced Gouldian you're most likely to see, there are two other varieties, one with a red face and the other (rarer) with a yellow face. Whichever one you see seems at the time to be the most stunning. This finch was first describd by the famous naturalist and artist John Gould in 1844. He named it after his wife, who had died relatively young. Gould wrote:

It was with feelings of the purest affection that I ventured to dedicate this lovely bird to the memory of my late wife, who for many years laboriously assisted me with her pencil, accompanied me to Australia, and cheerfully interested herself in all my pursuits.

The bird is called *Amadina gouldiae* and it seems to be a great pity that Gould didn't use his wife's own name, Elizabeth, rather than his surname as the specific epithet. Because nowadays, with little learning of Latin, who but a biologist would know that the 'ae' at the end of *gouldiae* makes it a feminine word? It could mistakenly be thought that Gould was commemorating himself rather than his wife—except that it's beyond the scientific pale to name a species after yourself.

Could I be wrong in thinking that the genus name *Amadina* also owes something to affection? Perhaps it comes from the Latin *amare,* 'to love'. If so it was by a happy accident, because that part of the name was not of Gould's own choosing. The genus name has been changed several times since his time.

Gouldian finches make their nests deep inside tree holes. Parent birds with food for the chicks must go into darkness out of bright sunlight, so, half blinded, how do they locate the open beaks of their importunate offspring? Astonishingly those gaping beaks wear a coloured luminous spot on either side, an accurate guide for a parental beakful of goodies! I wonder if this was known in Gould's time, when there was no clever and intrepid cameraman around, no David Attenborough to bring such marvels to our living rooms.

While our Australian finches may not be very familiar to us in their wild state, in person as it were, most are well known here and overseas as cage and aviary birds. Some have been successfully kept and bred in Europe and America for more than a hundred years.

It's an interesting thought that around the same time that our native finches were on their way to Europe, there were three European finches on their way to Australia. In a boat, I mean. The finches were imported, together with thrushes, blackbirds and skylarks, by the Victorian Acclimatisation Society, which released them to add a touch of home to a countryside still very alien-seeming.

Two of those European finches, the Goldfinch and the Greenfinch, are still holding their own in the southern states. And the other? Widespread and doing very nicely, thank you. It's our friendly House Sparrow, which is a true finch although it seems always to be in a class of its own.

'A charm of sparrows?' Yes, I think so, in both senses of the word, in spite of the nuisance they can be. After all, sparrows have been around picking up our crumbs for probably thousands of years.

Who knows, if we're still around on this planet a few thousand years from now our own native finches might have become just as familiar around our homes as their House Sparrow cousins. Perhaps even the gorgeous Gouldian.

Less spectacular than some, but full of personality—the Double-barred Finch (Poephila bichenovii).

Keeping your cool on a desert dune

If you ever feel the urge to run up an outback sand dune without your shoes on, don't! The first time I ever got close to one of those smooth red slopes I was seized with a irresistible urge to run up it barefoot. I was halfway up before my feet got their scream of protest through to my brain. You know how long it takes when you stub your toe.

Instinct said 'shuffle' so I shuffled. Ankle deep, the sand was at least 20 degrees cooler. So I shuffled ankle deep all the way back down the dune. There were blisters on my soles next day.

Well, it taught me a lesson about life in the sandy deserts of the arid zone. No wonder most of the small animals out there stay underground all day and do their living at night. The proof is there to be seen on any outback dune at dawn, written in the sand. Tracks and trails of every sort.

Only a few creatures venture out by day and those mostly at early morning and late afternoon. But there are some that brave the searing heat from dawn to dark, and that's what cameraman Jim Frazier and I had to do for several weeks while filming the wildlife of a central Australian sand dune for a television documentary.

Most activities on the dune were centred around the scattered clumps of spinifex or porcupine grass. These spiky thickets not only provide food and protection from predators, but also make islands of coolness in a sea of sand hot enough to keep even the ants underground. But we noticed the covered highways of the remarkable little Spinifex Ants linking some of the spinifex clumps.

Outback Spinifex Ants build shelters over insects that give them honeydew.

Spinifex Ants may look like common or garden 'little black ants' but they live only in the outback. The special thing about them is their close dependence on a particular kind of spinifex and the use they make of the sticky resin it produces.

The ants live in underground nests right in the middle of the spinifex clumps. To make the nest they gather the soft resin drop by drop from the spinifex leaves and mix it with sand. Like cement, the resin sets firm. The same mixture is used to seal and roof over the permanent runways between the ants' home base and other spinifex clumps, allowing them to move with impunity over the hot sand.

ABOVE: *By building covered runways between the spinifex* (Triodia) *clumps, the ants avoid burning their feet.*

But the most intriguing use of the resin mixture is in the building of shelters over clusters of scale insects that feed on the spinifex. The ants 'milk' these insects for their honeydew. The shelters built around the exposed spinifex stems allow the ants to visit their charges right through the heat of the day. And it's quite

likely that the scale insects would perish of desiccation without their protection.

Ants of various kinds are the main food for dragons on the dunes and one of our briefs was to find and film that heraldic-looking little desert dragon *Moloch horridus* demolishing a column of ants. You don't come across one of these Thorny Devils every day, but we were lucky enough to find one not far from the dune. The problem was to find a column of ants out in the bright light of day. The Spinifex Ants were too well protected in their tunnels.

There was one other little ant that could be relied on to form exposed columns along well-worn trails. But life is never easy for the wildlife

LEFT: *Tiny Thorny Devils* (Moloch horridus) *newly hatched beneath the sand of a central Australian sandhill.*

With its spiny coat and strange hump, the Thorny Devil is one of the world's most bizarre reptiles.

cameraman; these ants came out to forage only in the late afternoon, when shadows were creeping across the dune. And for filming the dragon and the ants close up Jim needed strong light.

We found a solution. I fooled the ants by holding an umbrella over both nest entrance and runway well before the ants normally emerged, and whipping it away as soon as they were out and running. Our Moloch, set down beside the trail, went to work on the outriders immediately, zapping them up with his tongue so fast the eye missed the action. By the time the ants got hot feet and raced for home we had our shots.

There were a few other insects to be found by day, with difficulty as most of them were coloured and shaped to blend with the spinifex. The stick insects and striped katydids fed on the spinifex. The Praying Mantises fed on the stick insects and striped katydids.

But the most amazing example of camouflage was a small grasshopper stippled to match exactly the multi-coloured sand grains. At rest on the sand, a fringe of hairs along the body concealed its shadow. For further security it had only to do a quick shuffle and kick a little sand over its back and, hey presto! no grasshopper. All that remained visible, if you had followed the action so far, was a pair of eyes that looked like two extra-large sand grains, and a pair of antennae like tiny sticks. This survival ploy seemed to work

JIM FRAZIER

ABOVE: *In the heat of the day a Military Dragon* (Amphibolurus isolepis) *ambushes insects from the shade of spinifex clumps.*

TOP: *Ants abound in the desert and the Devil feeds on nothing else.*

well; next to the ants these were the most common animals around the dune.

A Thorny Devil's tongue is the fastest thing about it. No need for speed with protective armour-plating and food provided on a conveyor belt. By contrast the long-tailed Military Dragons that share the same habitat seem almost jet-propelled. One of these, a male holding a territory on our dune, was surprisingly tolerant of our daily presence.

Out on the dune all day, this dragon spent his time resting under the spinifex or moving about from one patch of shade to another. If he ventured onto the hot sand it was for an occasional dash after a stray ant or a fly or even a bit of debris blowing in the wind. Late in the day when the ants came out he would settle near one of their trails for a proper meal. I think most of the time he was just 'showing the flag' as there were continual border skirmishes with neighbouring dragons. Every time this happened, we lost him . . . his speed across the sand between and around obstacles was unbelievably fast. There was no fight, just a brief confrontation before the intruder fled.

There's always a special thrill when a wild animal volunteers trust in a human. With his remarkable eyesight the dragon must have spotted our daily arrival from far off, yet we always found him carrying on as usual. It wasn't long before he was hunting ants between the tripod legs, and with care we could bend down and scratch his head. In the proprietorial way of humans we came to think of him as 'our' dragon, looking forward to our daily rendezvous on the dune. Military Dragons were not written into the script, but we soon changed that.

For wildlife film-makers, time at any one location seems to run out just when you've settled down to a comfortable relationship with the local inhabitants. Now, once again, and all too soon for Jim and for me, it was up tripods and away. We had an appointment to keep with some termites in Arnhem Land.

26 Colourful characters, slugs

The more I learn about slugs and snails the more amazing I find them. The basic facts are impressive enough. Imagine having your teeth on your tongue, and thousands of them at that. Imagine having your sexual apparatus on one side of your head. Yes, there's more to gastropods than crunchy shells and slimy trails and mayhem in the lettuce patch.

Speaking of shells, it's often thought that a slug is just a primitive sort of snail that's never had the sense to grow one. But it's really the other way round. Slugs are wise guys. Slugs have been there, done that, tried the shell experience. Got tired of carrying their houses around on their backs and instead opted for mobility.

So it's tempting to define slugs as snails that have discarded their shells. But this isn't quite right, either, and here's another surprise: many slugs still have a shell, even if it's only a rudimentary one tucked away out of sight. The introduced slug *Testacella haliotidea,* for instance, has a small, disc-like shell sitting rather comically right on its rear end, like a cap about to fall off.

Then there are the Helicarions, the little semi-slugs you find feeding on some rainforest fungi. These are snails with a delicate shell worn on the middle of the back like a saddle, too shallow to retreat into completely and often covered by the attractively patterned mantle flaps. Helicarions are thought to be well on the way to slugdom.

Let me explain about those mantle flaps. The mantle is a fleshy covering or fold, often with lobes or flaps, that all slugs and snails have on the upper part of the body near the front. In snails it's the mantle that secretes the shell and has other useful functions. In slugs the mantle may cover most of the body or be so small as to be unnoticeable. Mantles vary so much they're a good clue to identification, as shells are for snails.

It must be said against slugs—and this applies only to earthbound ones; their marine cousins are another thing again—that they mostly come in drab colours of brown or grey or at best yellowish. Even the introduced Leopard Slug with its striking pattern of stripes and spots is sombrely dressed. But now to a slug of a different colour. And probably a different flavour, too, if nature's use of red as a warning to vertebrate predators holds good for gastropods.

Triboniophorus graeffei is a native Australian slug found along the east coast and Great Dividing Range, from Sydney's heathland and dry sclerophyll to the tropical rainforest on top of Thornton Peak in north Queensland. What a surprising variety of habitats for a single species! And the other surprising thing is the number of colour forms it comes in.

On damp or rainy nights I can usually find one or two of these slugs feeding on the sandstone outcrops around my Sydney garden. Although they're big in gastropod terms (12 centimetres long or more) the slugs are hard to see because they match the colour of wet sandstone and blend so well with the shapes of fallen gumleaves.

I look for a leaf shape with a thin scarlet border all around and a small triangular patch outlined in scarlet at the broad end. The skin has an overall pattern of 'veins' rather like a leaf, and a dense sprinkling of small tubercles that gives it a bubbly texture. Just inside the scarlet triangle, the breathing hole or *pneumostome* opens and shuts slowly as the slug breathes.

In other parts of Sydney there's a cream or pinkish-grey form of the same species with a

A vestigial shell shows that Helicarion snails are well along the evolutionary road to slugdom.

OPPOSITE: *One of a number of colour forms of Australia's largest slug* (Triboniophorus graeffei).

BOTTOM: *A colour form from the New England district of New South Wales.*

TOP: *The red slugs are known only from the Narrabri district of northern New South Wales.*

ABOVE: *The feeding tracks of* T. graeffei *on Sydney's bluegums are more familiar than the slug itself.*

much smoother skin. It matches the smooth-barked bluegum trunks the slugs are often found on.

The name *Triboniophorus* means literally 'cloak-bearer', but in fact their mantles are very small, just the triangular area that's outlined in red. It is this colour pattern that has given Sydney forms of *T. graeffei* the common name of Red Triangle Slug. To the north the slugs are much more highly coloured. They may be rose pink all over, or yellow with a solid scarlet triangle, or plain pillarbox red.

Triboniophorus is a fungus feeder, but it doesn't eat mushrooms, toadstools, bracket fungus or any of the more obvious kinds. Like a cow grazing in a grassy meadow, the slug grazes on the crops of minute fungi that appear so quickly on rocks and tree trunks in damp weather. Too small to see individually, these fungal growths are sometimes visible as a kind of surface bloom. But their presence is best revealed by the tell-tale signs left behind by a feeding slug.

Have you ever seen a set of scalloped tracks criss-crossing the face of a rock or patterning the trunk of a smooth-barked tree? This is the characteristic feeding track of *Triboniophorus* slugs. They eat while on the move, and each individual scallop marks a mouthful, scraped up by that remarkable rasp-like tongue called the radula.

A word about that radula. Slugs and snails don't have our kind of teeth. Instead they have a ribbon-like 'tongue' with rows and rows of highly specialised backward-pointing teeth all over it. According to their bearer's way of life, the teeth may cut and tear, or rasp, and they operate on the tip of the 'tongue' which is extended out from the mouth. There may be thousands of tiny teeth for scraping plant tissue or only a few big sharply pointed ones for tearing flesh. And as a radula continually grows forward, rows of worn teeth are replaced by new ones.

As with all native slugs and snails, *Triboniophorus* slugs are incapable of harming garden plants or crops. The harmful garden-type slugs and snails came with us to Australia, like the domestic cockroaches and other pests. Native slugs haven't the slightest interest in us. But our interest in them is both scientific and aesthetic. *Triboniophorus* belongs to a family of slugs called the Athorocophoridae. The family is confined to the Australasian region, and in many

ABOVE: *The slug feeds on minute plant growth on tree trunks and leaf litter, doing no harm to garden plants.*

LEFT: *Helicarion semi-slugs feed on fungi in damp forests.*

ways its members are unique. For one thing, they have a rather more advanced kind of lung than other land-living molluscs. For another, they have only one pair of tentacles instead of the usual two pairs and this is a feature that makes for easy identification of the group.

But to non-scientific eyes the outstanding thing about the *Triboniophorus* slugs is their attractive colouring. For that alone, even the most ardent gastrophobe among us might admit to their charm and forgive them their sluggish ways.

27

Child labour in the weaving industry

Imagine the scene. It's 1770 and Captain Cook's sailors are relaxing at last after that epic voyage north through perilous, uncharted reefs of the north-east coast.

'. . .gripping, nipping and squirting. . .' The defence strategy of the Australian Weaver Ant or Green Tree Ant (Oecophylla smaragdina virescens).

Imagine them stretched at ease on the golden sands of a far north Queensland beach, the chatter and screech of exotic birds sounding from the dense rainforest behind them, the casuarinas sighing in the wind.

Imagine them all of a sudden leaping to their feet with cries of dismay, clutching their heads, tearing their clothes off, racing in panic down the beach to roll naked in the warm, shallow waters ...

What madness could have affected them? What venomous horror has come creeping out of the forest in this unknown land to attack and perhaps kill them?

Well, that's the way some present-day visitors to tropical Queensland react, too, the first time they're attacked by a colony of *Oecophylla smaragdina virescens,* Green Ants, Green Tree Ants, Australian Weaver Ants. Crawling in your hair, biting your toes, sliding down inside your shirt. Masses of little green and gold bodies, soft yet oddly tenacious of life, gripping and nipping and squirting wherever they can get a foothold.

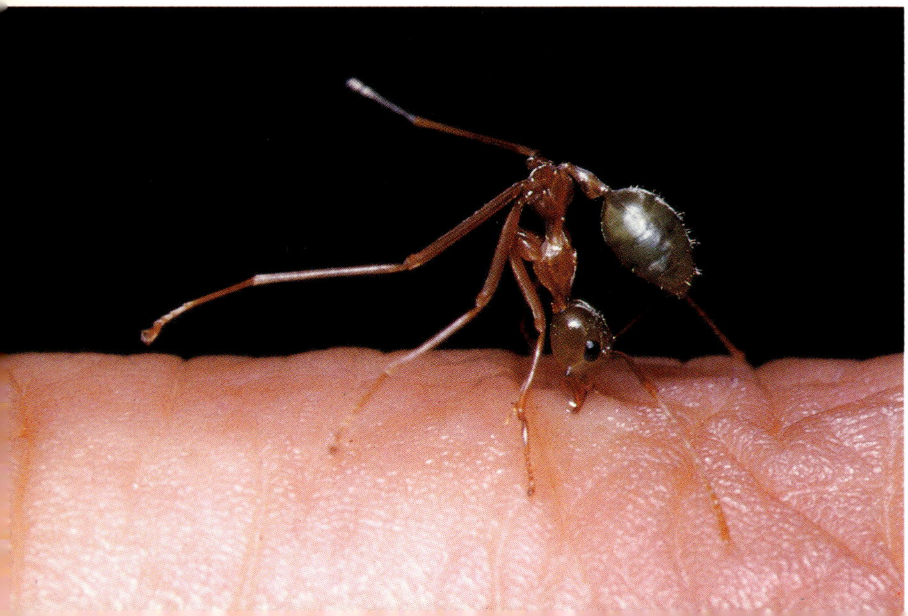

But not stinging. Weaver ants don't have stings, they bite with their mandibles. And as they bite with one end they squirt an irritant chemical out of the other end.

After a while you come to realise they're not hurting very much after all. And anyone from the south who's been attacked by giant Bulldog Ants will wonder what all the fuss is about. No, it's mostly psychological; there are so many of them, they're so small, and they seem to be raining out of the trees all over you. Of course you find eventually that there weren't hundreds at all, just a half-dozen or so squashed dead bodies, sacrificed for home and queen.

After a while you learn to recognise and avoid their leafy nests in the bushes and trees along rainforest tracks. And on the beach you learn to put your bare feet down carefully between those long succulent stems that creep across the sand above the high tideline. These stems are the ants' causeways over the loose sand, highways with traffic often bumper to bumper as supplies are delivered to isolated nests, special repairer ants brought in for reconstruction work, or colonies moved holus-bolus from tree to tree.

And after a while you find yourself stopping to watch, because there are few insects as fascinating in their behaviour as ants, and as these ants live above ground they're easier to study than most others.

Weaver Ants occur right across the top of Australia and some distance down the tropical Queensland coast. Northerners call them Green Ants or Green Tree Ants. Their relatives in other parts of the world are generally known as Weaver Ants, or Red Ants, though some are more brownish-yellow than red, and not as pretty as ours. It is their ability to weave that makes all these ants famous, including our Australian ones, and I think that merely calling them Green Ants hardly does justice to their remarkably sophisticated behaviour.

As ants go these are attractive little insects with their green and golden colouring, and their scientific name is also attractive. *Oecophylla smaragdina*

Worker ants pull leaves together in the first stage of making a nest.

If the leaves are far apart, workers form a living chain to draw them closer.

means 'little emerald that lives in a leaf-house'. The extra word *virescens* tacked on the end of the Australian species means 'becoming green', which is a little puzzling as the adults are always green.

The ants make their nests out of the living leaves of trees and shrubs. The nests are rounded structures often as big as a football. When the leaves are small there's a lot of work involved in joining them all together and filling the spaces between with a fabric of white silk. One colony of ants can occupy a large territory with a number of nests in adjacent trees. The queen ant lives hidden away inside the nest. Inside, too, are the brood chambers, the nurseries where workers take care of the eggs and creamy-coloured grubs (larvae) and pupae.

All the ants you see about in the open are workers, big ones and small ones—major and minor. The big ones forage for food and defend the nest fiercely, standing up high on their legs, shaking and jerking with a great show of aggression, their mandibles spread wide open. Nest-making, the joining of leaves with silk, is a cooperative effort by workers of both sizes, each with her specific role.

Question is, where do they get their silk? Adult ants have no way of secreting silk, no silk glands. But their larvae do, and the astonishing thing is that the adult ant workers use those little sisters, the ant grubs from the nurseries, as silk shuttles. This is the sequence of events.

When a nest is to be made, the first thing to do is draw the edges of adjacent leaves close together. In a remarkable act of coordination, large workers combine to link the leaf edges, bridging the gaps with their bodies and pulling. If the gap is a big one they form a living chain across it.

As soon as the leaf edges are close enough, a small worker appears on the scene with one of the grubs in her mandibles. When she reaches the site she dabs the head end of her charge against each leaf edge in turn, laying down a thread of silk across the gap. In turn others come with their 'shuttles', some working from the inside, some from the outside. As the silk dries it shrinks, pulling the leaves even closer. Now the stretched-out workers can relax their hold and start again somewhere else.

As soon as a grub runs out of silk it is returned to the nursery and replaced. And so it goes on, until the nest reaches its optimum size. The grubs of Weaver Ants can never make cocoons for themselves. Instead they sacrifice their silk to the good of the colony. But that doesn't stop them turning into perfectly normal, healthy workers, like their older sisters.

There are many other remarkable things to know about these ants. For instance, they use their weaving skills to build shelters around their 'cattle'. These can be seen as miniature nests of silk and leaves on stems where sedentary

Worker ants use their little sisters, the larvae, as shuttles to weave leaves together. Only the larvae can produce silk.

ABOVE: *Certain butterfly caterpillars provide the ants with honeydew in return for protection.*

TOP: *Silk shelters or byres are woven over the ants' 'cattle', the sap-sucking insects that provide them with honeydew.*

sap-sucking insects feed. Look inside and you'll see a few worker ants milling about around a colony of scale insects or some frog-hoppers or aphids. These insects come to no harm from the ants, in fact quite the reverse. In return for 'honeydew', the sweet substance they excrete, the ants protect and shelter their charges.

Then there are certain caterpillars of the family of 'Blue' butterflies (Lycaeenidae) that have a special relationship with these ants, and probably can't live without them. Such a caterpillar has special glands on its back that turn it into a kind of

LEFT: *The ant will 'tether' its large victim until other workers arrive to help.*

mobile vending machine. When an ant stimulates a gland with her antennae a little window opens and out pops a glistening drop of sweet fluid. The ant sucks it up and the window closes. The ant may ride these caterpillars about the food plant, going with them from their silk shelters to the leafy stems where the caterpillars feed and back again. Whether it was designed that way or not the ants must certainly be a deterrent to the many predators that regard caterpillars as a highly desirable meal.

In spite of their care towards their special charges, the Weaver Ants are themselves fierce predators, cooperating as a team to capture living prey for the carnivorous larvae back in the nest. A single ant will seize part of an insect even as big and fierce as a wasp. It will hang on to a leg or antenna, using its own spread legs and body as a tether until more workers come along to help it. Then the prey is properly subdued and carried home to the nurseries.

There are other relationships between Weaver Ants and their neighbours that are more puzzling. Certain little jumping spiders with iridescent colours always seem to be be around the ants' nests and in particular around their 'cattle shelters'. Do they prey on the ants? Do the ants prey on them? There seems to be a truce between them but the spiders must be there for some reason. This is one of the mysteries of the Weaver Ant world still waiting to be solved.

Worker ants co-operate to repair a damaged nest.

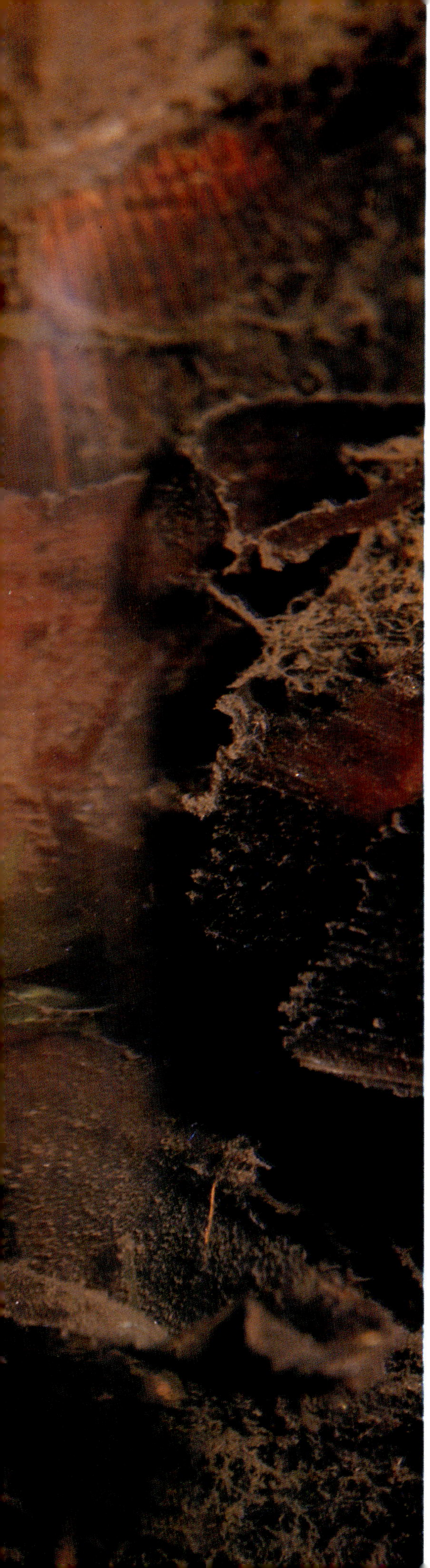

28 Prothalamion in a garden pond

'Toc!' Legs dangling, the big striped frog floats upright at the edge of my garden pond. The pale bubble of his inflated throat gleams in the dark like a life jacket supporting his head above water. One fat hand rests on a floating leaf, the other against the sandstone coping. Blank eyes protrude, reflecting twin moons.

'Toc!' The call is explosive, and the fractional recoil of the frog's body rings the water around him with ripples. 'Toc! ... Toc! ... Toc!' He calls with his mouth closed and his vocal sac stays inflated between calls.

Around the pond a dozen or more of his kind are calling from the water. Single notes distinctively pitched according to the age and size of the caller follow in sequence to produce a simple, random tune. The frogs are males of a common Australian species, *Limnodynastes peroni*. Guarding the position they have reserved for the night, they make their whereabouts known, calling the females from their hiding places around the garden.

A few feet from the frog I'm watching and at right angles to him a female has been floating for some time. Submerged like a miniature hippo with only eyes and nostrils above water, she gazes with seeming indifference in front of her. Suddenly she darts towards the male and comes to rest 15 centimetres away, with her back to him.

Immediately the quality of the male's call changes. Up to now he has been taking part in a carefully spaced call sequence, in which the notes of individual frogs are now and then transposed, rather like a company of bellringers ringing the changes. Sometimes two males will 'pair off' vocally and make their calls simultaneously, perhaps in the interests of clarity. The pond is overcrowded and a random cacophony of blurred notes might confuse the females.

But this male now drops out of the choir to concentrate on the female who has, for no doubt good reasons of her own, picked him out of the

Clasped by the male, the female Striped Marsh Frog (Limnodynastes peroni) *beats up a protective froth around her eggs.*

The male frog's expanded vocal sac acts as a resonating chamber for his monotonous calls (L. peroni).

line-up. His call becomes a gentle entreaty, a clucking come-hither to which the female bats no obvious eyelid . . .

Another frog surfaces at the female's side with a swirl of disturbed water and she hurries off, alarmed. The challenger adopts an upright stance near the territory-holder, inflates his throat and 'tocs' loudly. Immediately the first frog deflates his throat, swings around and approaches the intruder, legs spread behind him. He utters a hoarse groan.

The stranger groans in return and lunges and the two wrestle in slippery combat, thrusting shoulder against shoulder. One dives under the overhanging rocks and the other follows. They merge, pressing throat to throat, chin to chin, snouts pointing skyward, limbs ineffectually grappling for a hold. Now they pause, gazing over each other's shoulder as though indifferent, then break their hold to resume the fight, groaning and clucking the while, finally diving out of sight together into the deep water.

The female, back again, goes on staring at the wall, waiting . . .

One of the two males surfaces. In a flash he's back in position, almost doing a back flip in his hurry to inflate his throat and resume the earlier rhythmic calling. He's identifiable to me as the original territory-holder by a graze on his head. He appears unharmed by his fight; the call of his rival can be heard now from a safe distance. The female moves back to his side. Enough of this love talk, she means business. Sharply she nudges the male in the shoulder with her snout and turns to lie alongside him, pressed against his flank.

The male is clucking quietly and continuously now as he gently manoeuvres behind and above the female, resting his throat on her head, grasping her slippery body tightly with his arms encircling her loins. This is the spawning position that enables him to fertilise her eggs as she lays them into the water.

Around the engrossed pair the racketing concert goes on, but barely audible. Behind it there's now another sound. It's a faint, rhythmic splashing coming from the far side of the pond where a mass of white foam floats against the dark sandstone coping.

At the edge of the foam two dark heads can be seen, the smaller on the surface of the water, the other above and behind it. This is another pair of frogs which has been amplexing for some time. *In amplexus* is the special term used for frogs in the spawning position; it simply means 'embracing'. In front of the embracing frogs there's a cluster of tiny bubbles on the water surface. A chain of bubbles extends backward on either side past the male's head to join up the gelatinous froth that partly envelops them, a froth speckled with tiny black eggs.

The splashing sound comes from the female. She is beating the water with her hands, never breaking the surface, but producing a rhythmic 'slop, slop, slop' of sound. She uses each hand alternately, thrusting outward and downward, pausing between each set of four beats. Her head reacts with a slight dodging motion. After every

LEFT: *The eggs are supported and protected by the jelly-like froth produced by the female.*

fourth beat the male moves forward a little in reaction then uses his hind legs to regain his position.

The female is laying her eggs and at the same time releasing a substance that can be whipped to a froth. Meanwhile the male fertilises the eggs as they emerge. The fertilised eggs will float in their tough, protective, jelly-like foam until they hatch.

For two hours this pair has been in position. Every minute the female completes 19 sets of four movements. In two hours she has made 9120 separate movements with her hands. No wonder she looks a little desperate!

The egg mass would be somewhat larger by now if it were not already nourishing some other pond dwellers. A smacking sound comes from below, and red bodies flash intermittently as the goldfishes suck in their omelette between pouting lips. By morning there will be little spawn left. By tomorrow evening, love's labour will be almost entirely lost.

Tadpoles that succeed in hatching are preyed on by fellow pond-dwellers such as large water bugs and beetles, and the fishing spider Dolomedes that dives below to catch them. They can also be taken by wading birds.

For frogs the major enemies are the snakes that hunt them both in the water and on land. But for the frogs in my garden pond the greatest danger comes from a more insiduous predator, one highly specialised in the techniques of frog-hunting.

A tattered waterlily leaf lying just below the surface of the pond suddenly comes to life. In front of it, what appears to be a half-submerged, shiny brown lily bud blinks open golden-ringed eyes set behind gaping nostrils. The golden rings glance about purposefully. Now the head is withdrawn and moves slowly forward below the surface, dragging behind it on the end of a long neck what had seemed to be an old lily pad. It is the carapace of the Long-necked or Snake Tortoise, *Chelodina longicollis.*

All through the summer the tortoise has basked in the sun close to the surface of the pond and his carapace carries a growth of green algae trailing behind him like hair. Only at night does he break his perfect camouflage to quarter the pond, revealing himself as an agile and ruthless predator.

Now, neck stretched fully and head questing, he paddles slowly around the pond. Unmated frogs are still calling. Close to one of them the tortoise stops. Does he see with those golden eyes, or is the scent of the frog carried through the water to his pit-like nostrils?

The tortoise points like a gun dog then moves forward, hardly stirring the fallen leaves that float above him or the stems of water weed through which he manoeuvres. His head stops a few centimetres from the frog but his body continues its forward movement until the long neck is bent sideways in a loop. The frog still calls, but hesitantly now.

In a flash the tortoise uncoils its neck to the fullest to lunge at the frog from below. He misses. The frog has shot from the water like a squeezed lemon pip. He crouches on a lily pad, his deflated throat pulsing rapidly. The tortoise noses around a little, butting his snout against the rock wall where the frog has been, then goes on his way.

Now he reaches the spawning pair and he stops, casting about with his head to pinpoint their position. The frogs are intent and unaware, the female's egg-beating continues unabated. The male is barely visible now under the rising mass of foam.

The tortoise moves in for the strike. He draws his body close to his purposefully curving neck . . . still the little female doggedly beats the water, the clasping male gazes straight ahead . . . and with a flurry of foam and water the snapping jaws have grabbed their prey. The female, hampered by the male, has been seized from behind. Oddly submissive, perhaps paralysed, with arms and legs extended she is borne to the bottom of the pond in a steep dive, her white belly and throat shining clearly up through the water in the grasp of that snake-like head.

Once down there the tortoise starts a violent flailing movement with his long neck, swinging his head from side to side. At each swing the powerful claws on alternate sides rip and tear at the frog's soft flesh. The frog is too big to swallow whole. The tortoise must use its jaws and claws and the powerful whiplash of its neck to rip its victim into manageable portions. Now, still lashing and tearing, he has moved to the deep end of the pond to finish the job hidden by the murky water.

The male frog has disappeared. The goldfish disturbed at their feast have moved across the pond to where foam is gathering now behind the newly amplexed pair of frogs. Already they hang beneath it, quivering, red-gold, smacking their lips and doing their bit to preserve the balance of nature in the small world of a lily pond.

JIM FRAZIER

Paterson's Curse or Salvation Jane? Whatever you call it, this introduced weed undoubtedly adds beauty to the landscape.

29 It's all sunshine and flowers for an outback lizard

To the unprejudiced eye the swathes of royal purple that have come to colour our rural landscapes in recent decades are a beauty and a joy to behold.

To the prejudiced, though, the flowers of Paterson's Curse are an eyesore, stealing good land from the pasture grasses grown to fatten sheep and cattle for the slaughter. Well, they're all intruders in the land, the sheep and cattle, the intro-

LEFT: *Shingleback Lizards are widespread across the continent and come in several colour forms* (Trachydosaurus rugosus).

ABOVE AND RIGHT: *It's all bluff, those tiny teeth can do no more than grip—unless you're an insect or a flower.*

duced pasture grass and the vagrant Paterson's Curse. And I know which of them I prefer to see.

In South Australia the plant is more kindly called Salvation Jane. It has survived drought and saved stock when foreign pasture grasses have perished. Bee-keepers love it because it makes good honey. In Europe, of course, *Echium* is not a weed but a wildflower, and I suppose the distinction between the two terms depends a lot on your vocation in life.

One native Australian has no equivocal thoughts about the matter. That anthophagous outback reptile the Shingleback Lizard finds the purple blooms of Salvation Jane an attractive alternative to native flowers. On location in far western New South Wales recently, we were surprised to find Salvation Jane in a true outback setting, the purple blending harmoniously with that true-blue plant of the outback, the bluebush. I saw kangaroos standing knee-deep in flowers and I must say they looked most exotic.

Whether or not kangaroos eat the flowers of Salvation Jane, the Shingleback certainly does. We saw one doing it beside a warm-red dirt road; there could be few sights less congruous than this armour-plated reptile nibbling the petals of a wayside flower.

In this strange, upside-down land of paradox it would have taken little to surprise early explorers. Even before they left their safe home for the great down-under adventure, reports from sea-faring men of strange animals glimpsed on brief visits ashore must surely have left the imagination wide open.

Such enigmas as the furry, duck-billed, egg-laying, web-footed platypus were to come later. At the turn of the 17th century a lizard with two heads (one at each end) was enough to go on with. And that was the first impression given by the harmless, sleepy-looking Shingleback Lizard. It's a first impression shared by many outback travellers today. The gentleman pirate William Dampier first recorded the lizard's strange appearance in his journals. He called it 'a sort of Guano's' and described it as follows:

Stump of a Tail that appears like another Head; but not really such, being without Mouth or Eyes. Yet this Creature seem'd by this Means to have a Head at each End; and . . .the Legs also seem'd all 4 of them to be Fore-legs . . . as if they were to go indifferently either Head or Tail foremost.

More than anything, Dampier seemed concerned with the lizard's potential for the cooking pot. I suppose that was important on a long ocean voyage, with any kind of fresh meat an essential addition to shipboard fare. Never mind the lovable characteristics of the thing, what sort of stew will it make?

In the outback, flowers of Salvation Jane are among the favourite snacks of Shingleback Lizards.

Dampier refers to other kinds of 'Guano's' (using this form of iguana as a term for lizards generally) as being good eating. Then he goes on to say:

. . . if pressed by Hunger, yet I think my Stomach would scarce have serv'd to venture upon these N. Holland Guano's, both the Looks and the Smell of them being so offensive.

He must have been referring to the smell when cut open, perhaps the gut contents. I've never found live ones at all offensive.

Always, for these first visitors in a strange land, the first thought on encountering a potential meal must have been: 'Does it bite? And if so, is it venomous?' As regards biting, a first-time encounter with a Shingleback could have been quite alarming, as it still can be. Even when we know it to be a gentle, inoffensive lover of flowers and sunshine. The suddenly gaping mouth with bright pink interior and blue protruding tongue are deterrent enough. But it's mostly bluff. Dampier again:

When a Man comes nigh them they will stand still and hiss, not endeavouring to get away.

What else can they do? Those little legs are not built for speed. Instead the Shingleback carries its protection in a tough, horny covering of scales that overlap like roof tiles. The Shingleback has aptly been described as an animated pine cone.

Shinglebacks have other names. Stumpytail, Bobtail, Scalyback, Sleepy Lizard and Bob-eye or Bog-eye, the origin of the last two names being obscure. They're found across southern and central Australia in one colour form or another. In the eastern states Shinglebacks are thought of as outback reptiles, as they don't live anywhere near the coast. In fact there's great difficulty in keeping them alive in the humid conditions of the east coast.

Shinglebacks are protected, with a licence required for captive specimens. But it's easy enough to see them in the wild on a trip to the outback, sunning themselves, snatching up insects in slow motion, or reaching up to snip the head off a yellow daisy or grab a mouthful of Salvation Jane.

Courting Shinglebacks are a common sight on outback roadsides

30 Co-existence in Australia's wild, wild city

'There's a dragon at the door!'

It's hardly the kind of casual remark you'd expect to hear from a suburban lady-of-the-house over lunch. Of course, what's at the door is not your run-of-the-mill, fire-breathing-type dragon. But it is an animal almost as unlikely as that mythical monster to demand entry to your premises for a handout of peanuts. It's that usually shy reptile the Eastern Water Dragon (Physignatus lesueurii) and in this house it's been part of the domestic scene for years.

A male water dragon and his harem share a suburban garden with humans (Physignathus lesueurii).

GLEN CARRUTHERS

When you enter the house the living-room appears to be dominated at first sight by a traditional Chinese painting of dragons around a waterfall. Look again! Those dragons, that rocky waterfall, are no picture, they're real, and they're just a peanut's throw beyond the living-room window, in the garden. Out there more than greens are grown in the vegetable patch. It's a sunny nursery where Ruth senior, female dragon, buries her eggs. In due season her hatchlings skitter out of the soil to safety in a vegetable forest.

Don't confuse these handsome reptiles with anything like the little garden skinks that sun themselves on our paths and patios. Fifteen-year-old Ralph measures a metre from arrogant snout to tapered tail tip and he's still growing. Ralph is the dominant male and in season he displays his status with a flush of dark red on the skin of his chest. Then, on the tiny manicured lawn or just as often on the flat roof of the house, he challenges other males to ritual battle, watched by his harem of lady lizards.

Few of us have a chance to visit dragonland but there are many other wildlife experiences that all Sydney people share. One of these is the nightly exodus of thousands of Grey-headed Flying Foxes from a suburban gully. These intriguing mega-bats fan out all over Sydney for their evening meal of nectar, pollen and fruit. Giant fig trees around city parks attract large numbers of them in season.

From the Opera House at dusk you can watch the bats flying in across the harbour for a feast of native figs from the gigantic trees in Sydney's Domain and Botanic Gardens. The story is told of a magical moment during one of Dame Joan Sutherland's open-air concerts. Right on cue, as Dame Joan was singing an aria from *Die Fledermaus*, a real bat—one of the Greyheads—came flying low over the heads of the delighted audience to settle in a nearby fig tree.

Most of Sydney's flying foxes live together in a suburban gully, roosting in the treetops. On hot days lucky people with a grandstand view from their balconies can watch these handsome mammals using their huge, unlikely-looking wings to fan themselves and their clinging babies. A viewing platform gives local and overseas visitors a look at a unique urban wildlife attraction. It is directly due to local interest in this colony that flying foxes have been declared a protected animal in New South Wales.

For wildlife overtaken by the suburban sprawl there can be advantages. The flying foxes, for instance, find soft fruits a welcome substitute for the dwindling supplies of natural food as we cut down the native trees for suburban development. The orchardists around Sydney's perimeter are not exactly happy about this.

To a Brushtail Possum mother and son, my boarded-up fireplace is just a luxury version of the hollow tree where their wild cousins sleep. At night the possums survey the scene from the chimney pot before going out to forage. At dawn it's back down the chimney to snooze and snore the day away.

Then there's the 'marsupial mouse' (*Antechinus stuartii*) that came to live for a while in my kitchen drawer. Basically the drawer wasn't so different from the usual nest site high up in some rocky crevice in the bush. The difference

Greyheaded Flying Foxes from Sydney's unique suburban flying fox colony (Pteropus poliocephalus).

was in the use the little resident made of local materials for bedding. As well as a bundle of dry gumleaves brought in from outside she used what she found in the drawer. She chewed a hole in a plastic bag containing a new dishcloth. She tore a red and white paper napkin into small pieces and stuffed these, together with the gumleaves, into the bag with the discloth. There she nested.

Once inside, the little animal was very snug and completely hidden. She came and went irregularly, as it's usual for Antechinuses (which are not related to mice) to have more than one daytime hiding place. I knew when she was at home by the way the plastic bag quivered at the opening of the drawer.

LEFT: *A plastic bag and some paper serviettes make a snug nest in a kitchen drawer for this Marsupial Mouse* (Antechinus stuartii).

BELOW: *A suburban Ringtail Possum* (Pseudocheirus peregrinus), *injured by a cat, recovers with human help.*

And just occasionally a long nose and a beady eye appeared at the opening to check on me.

At night it's the spiky geckoes that come out to hunt around Sydney's bushland suburbs. But by day a different kind of reptile does its bit for conservation in suburban gardens. For the harmless, handsome Bluetongue Lizard (*Tiliqua scincoides*) the introduced snails that wreak havoc in the flower beds are an exotic culinary bonanza. Escargots-on-the-rocks, laid on in never-ending supply.

These big, slow-moving skinks have remarkable blue tongues which they poke out as they go along, testing the air for good things to eat and for predators. They spend a lot of time lying in the sun, and they're easily tamed. Young Bluetongues are produced live. The pet female of a schoolboy acquaintance has produced 18 offspring at one birth; an astonishing feat when their total bulk at birth appears to exceed that of their mother!

Around most Sydney suburbs encounters with some form of wildlife are common, and it's often the rescue of lost, abandoned or injured infants that first sparks interest in the animals that share our city. Like the young Ringtail Possum brought in by the family cat and nursed back to health to

When hollow trees are scarce, chimneys provide alternative accommodation for Brushtail Possums (Trichosurus vulpecula).

JIM FRAZIER

JIM FRAZIER

ABOVE: *A baby Noisy Miner* (Manorina melanocephala) *that fell on good times.*

LEFT: *Exotically coloured King Parrots (*Alisterius scapularis) *quickly become regular visitors to gardens.*

OPPOSITE TOP: *Nectar-feeding Rainbow Lorikeets* (Trichoglossus haemotodus) *find exotic garden shrubs a suitable alternative to flowering eucalypts.*

OPPOSITE BOTTOM: *Kookaburra families exploit human families in the nicest possible way* (Dacelo gigas).

become a child's pet. Like the baby Noisy Miner that toppled from its nest and fell on good times, with food and shelter and its own private swimming pool provided by humans, until it could fly away.

Sydney is lucky. The city is bordered on three sides by great national parks, and contains many bush reserves and parks both urban and suburban. By an accident of topography much of the bushland in steep gullies survived when suburban development took place on the ridges. Since native gardens are now fashionable, people build with minimum disturbance to trees and shrubs. And birdwatching is a pastime increasingly taken up by people of all ages.

There's no doubt that the native birds have become the most familiar and best loved of all our suburban wildlife. And what birds they are! Where else in the world would you see flocks of gaudy screeching Rainbow Lorikeets and family parties

OPPOSITE: *Sydneysiders delight in the daily visits of Rainbow Lorikeets to garden feeding stations.*

BELOW: *The Bluetongue Lizard* (Tiliqua scincoides) *is harmless to humans and eats garden snails.*

of green and scarlet King Parrots flying freely around houses and feeding at bird tables? Not to mention the kookaburras and currawongs and the rest that have thrown in their lot with us. You might say they exploit us, but they do it in the nicest possible way.

The invertebrate animals are the least known of our suburban wildlife. Yet they're the most numerous, the most able to adapt to human intervention and survive it. There are insects around us as strange as science fiction creatures, or as beautiful as flowers. There are spiders worth more than a shudder or cursory glance for the fascinating complexities of their lives and for their engineering skills. That shadow world that lies about us is open to inspection by anyone with patience and a magnifying glass.

In Australia's early days, to establish a city involved the destruction of native life. Now, though casualties still occur, 'progress' is being redefined in kinder terms—kinder to the plants and animals that might once have been thought to stand in its way. In rural Australia the old conflicts between man and nature are still carried on. In the city we are not in competition with nature for our livelihood. It could be that the suburbs will become the last refuge for some of the wildlife that has taken a chance on us and stayed around.

Index